Study Guide

Biology Unit 1
for CAPE®

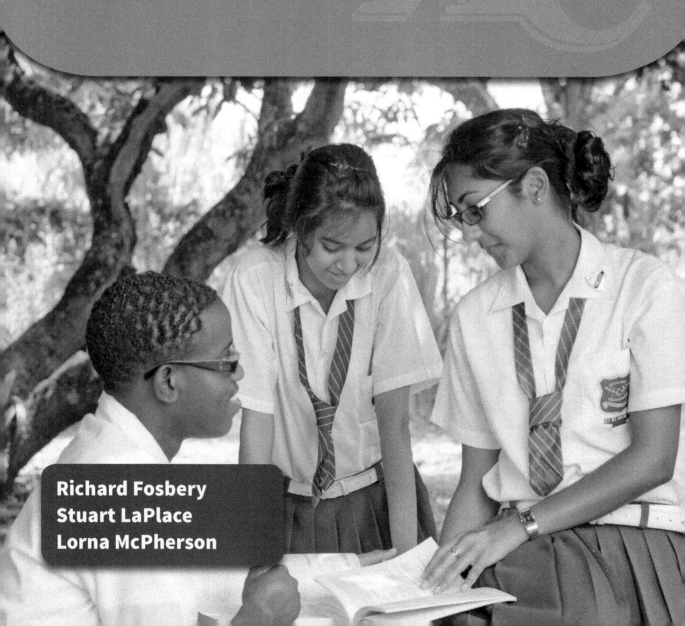

Richard Fosbery
Stuart LaPlace
Lorna McPherson

Great Clarendon Street, Oxford, OX2 6DP, United Kingdom

Oxford University Press is a department of the University of Oxford.
It furthers the University's objective of excellence in research, scholarship,
and education by publishing worldwide. Oxford is a registered trade mark of
Oxford University Press in the UK and in certain other countries

First published by Nelson Thornes Ltd in 2012
This edition published by Oxford University Press in 2015

British Library Cataloguing in Publication Data
Data available

978-1-4085-1646-1

26

Printed and bound by CPI Group (UK) Ltd, Croydon, CR0 4YY

Acknowledgements

Cover photograph: Mark Lyndersay, Lyndersay Digital, Trinidad. www.lyndersaydigital.com
Illustrations: Wearset Ltd, Boldon, Tyne & Wear
Page make-up: Wearset Ltd, Boldon, Tyne & Wear

The authors and the publisher would like to thank the following for permission to reproduce
material:

Photos
Module 1: 1.8.3 STEVE GSCHMEISSNER/SCIENCE PHOTO LIBRARY; 1.10.1 Laurence Wesson and John
Luttick (James Allen's Girls' School); 2.2.1 DON FAWCETT/SCIENCE PHOTO LIBRARY; 2.3.1 *Plant Cell
Biology on DVD, Information for students and a resource for teachers*, by B Gunning, Springer-Verlag 2009;
2.5.1 DR KEITH WHEELER/SCIENCE PHOTO LIBRARY; 2.8.1s author's own; 2.8.3a J.C. REVY, ISM/
SCIENCE PHOTO LIBRARY; 2.8.3b J.C. REVY, ISM/SCIENCE PHOTO LIBRARY.

Module 2: 2.2.1 STEVE GSCHMEISSNER/SCIENCE PHOTO LIBRARY; 2.4.1 GENE COX/SCIENCE PHOTO
LIBRARY; 2.7.2.i PR. G GIMENEZ-MARTIN/SCIENCE PHOTO LIBRARY; 2.7.2.ii PR. G GIMENEZ-MARTIN/
SCIENCE PHOTO LIBRARY; 2.7.3 CMEABG-LYON-1, ISM/SCIENCE PHOTO LIBRARY; 3.1.1 Glowimages/
Getty; 4.4.1 © Zmeel Photography/istockphoto.com; 5.5.2 Dr Claire Williams; 5.6.1a David Fox/Getty;
5.6.1b David Fox/Getty; 5.7.1 © Michael Stubblefield/istockphoto.com; 5.9.1 author's own.

Module 3: 1.1.1m © PHOTOTAKE Inc. / Alamy; 1.3.2 DR KEITH WHEELER/SCIENCE PHOTO LIBRARY;
1.3.4 IASPRR society; 1.4.1 Bruce Watson; 1.5.2 author's own; 2.1.2 JOHN BURBIDGE/SCIENCE PHOTO
LIBRARY; 2.2.2 POWER AND SYRED/SCIENCE PHOTO LIBRARY.

p30 Fig. 2.3.1 Plant Cell Biology on DVD, Information for students and a resource for teachers, by B
Gunning, Springer-Verlag 2009; p150 Fig. 1.4.1 Bruce Watson;

Although we have made every effort to trace and contact all
copyright holders before publication this has not been possible in all
cases. If notified, the publisher will rectify any errors or omissions at
the earliest opportunity.

Contents

Contents

Introduction

This Study Guide has been developed exclusively with the Caribbean Examinations Council (CXC®) to be used as an additional resource by candidates, both in and out of school, following the Caribbean Advanced Proficiency Examination (CAPE®) programme.

It has been prepared by a team with expertise in the CAPE® syllabus, teaching and examination. The contents are designed to support learning by providing tools to help you achieve your best in CAPE® Biology and the features included make it easier for you to master the key concepts and requirements of the syllabus. *Do remember to refer to your syllabus for full guidance on the course requirements and examination format!*

Inside this Study Guide is an interactive CD that includes electronic activities to assist you in developing good examination techniques:

- **On Your Marks** activities provide sample examination-style short answer and essay type questions, with example candidate answers and feedback from an examiner to show where answers could be improved. These activities will build your understanding, skill level and confidence in answering examination questions.
- **Test Yourself** activities are specifically designed to provide experience of multiple-choice examination questions and helpful feedback will refer you to sections inside the study guide so that you can revise problem areas.

This unique combination of focused syllabus content and interactive examination practice will provide you with invaluable support to help you reach your full potential in CAPE® Biology.

We have included lots of hints, explanations and suggestions in each of the sections. At the end of many of the chapters is a summary section that includes some activities to help assist your learning.

As you work through your CAPE® Biology course, read through any notes you took during your lessons. While doing this you should read textbooks, this guide and relevant up-to-date information from the internet. Use the information you find to add to your notes. In some places we have given you suggestions of searches you can make on the internet. Try to find good, accurate websites. Those that end in .edu or .ac are reliable. Entries in Wikipedia should always be double checked for accuracy.

When you finish a topic, answer the summary questions at the end of each section. You will notice that many of these start by asking for definitions of the terms relevant to each topic. This is to prompt you to use the glossary. At the end of each chapter are exam-style questions with advice on how to answer questions.

Try to find ways of summarising the information you have learnt so that you have some concise notes you can use for revision in the weeks before the exam. This is especially important for topics you have found difficult to learn and understand. You can write yourself bullet points about these topics. You can also make charts and posters to help your learning. There are many ways in which you can organise information and we give you some examples in this book.

1.1 Introduction to biochemistry

Chemistry of life

Just about everywhere you look the most common colour in your environment is green. It is the colour of chlorophyll, the main photosynthetic pigment in plants. It is used to absorb light and is involved in the conversion of light energy to chemical energy and as such is the basis of life on Earth. Chlorophyll is inside chloroplasts, which are organelles found within plant cells. Cellulose is the compound that forms the cell walls surrounding those plant cells. Cellulose is the most common biological compound on Earth.

Biochemistry is the study of biological molecules and their roles in organisms. These molecules are the building blocks of life – constantly being assembled and taken apart. **Metabolism** is the term given to all the chemical reactions that occur in organisms and can be divided into:

- **anabolism** – the reactions that build up larger biological molecules from smaller ones; they are known as anabolic reactions
- **catabolism** – the reactions that break down large biological molecules into smaller ones; they are known as catabolic reactions.

All the major compounds that make up organisms are based on the element carbon. This element can form covalent bonds with itself and with other atoms. These covalent bonds are very strong. It can also form single and double bonds. The range of functions carried out by biological compounds is staggering – just two of those functions are that chlorophyll is an energy transducer and cellulose forms fibres that are very strong. You will also learn that biological molecules provide energy, carry messages, catalyse reactions, store energy, store and retrieve information, transport gases and have many other functions.

The air around you is composed of oxygen, carbon dioxide, nitrogen, hydrogen and other gases. These are inorganic compounds. The complex compounds of carbon are known as organic compounds. Biochemistry is the chemistry of life and so we cannot progress without knowing some chemistry. Here are some terms you will have learnt but which you must know and use in your Biology course:

- chemical element – pure chemical substance consisting of one type of atom
- atom – smallest component of an element having the chemical properties of the element; the nucleus of an atom consists of neutrons and protons, and is surrounded by electrons
- isotope – atoms of the same element with different numbers of neutrons in the nucleus
- compound – a substance made from two or more chemical elements
- molecule – the smallest particle of a substance that retains the chemical and physical properties of the substance and is composed of two or more atoms
- ion – an atom or molecule or chemical group that has lost or gained one or more electrons and is either positively or negatively charged
- ionic bond – a chemical bond between two ions with opposite charges

■ covalent bond – a chemical bond involving the sharing of a pair of electrons between atoms in a molecule.

There are small molecules that are assembled into larger ones. In some cases these are assembled into much larger molecules. Glucose is a relatively small molecule that is assembled in different ways to give larger molecules. The small molecules are known as monomers, the larger molecules as polymers. Not all the large biological molecules are polymers. Lipids are made from a small number of sub-unit molecules.

The biological molecules are subdivided into groups, which are summarised in this table.

∞ Link

When asked to name the groups of biological molecules, students often forget nucleic acids. See page 62 for the structure and function of DNA and RNA.

Macromolecules	Elements	Sub-unit molecules	Examples	Roles in organisms
carbohydrates	C, H, O (ratio 1:2:1)	■ glucose	■ starch (amylose and amylopectin) ■ glycogen ■ cellulose	■ energy storage ■ support
lipids	C, H, O*	■ glycerol, fatty acid ■ phosphate (phospholipids)	■ triglycerides ■ phospholipids	■ energy storage ■ thermal insulation ■ electrical insulation ■ membranes
proteins	C, H, O, N, S	■ amino acids (20 different types)	■ haemoglobin ■ collagen ■ amylase ■ pepsin ■ insulin ■ antibodies	■ transport ■ support ■ catalysts ■ messengers ■ protection
nucleic acids	C, H, O, N, P	■ nucleotides (five different types)	■ DNA ■ RNA (mRNA, tRNA, rRNA)	■ information storage ■ information retrieval ■ production of proteins

* Note
Ratio of C:H = approx 1:2, but ratios of C:O and H:O are high, e.g. 9:1 and 18:1

Summary questions

Try the questions below and keep your answers in a folder alongside your notes.

1 Define the following terms:
 ■ organic compound
 ■ macromolecule
 ■ monomer
 ■ polymer.

2 List the four main groups of biological molecules.

3 Write down the chemical formula for glucose and show how to calculate its relative molecular mass.

4 Name one biological molecule that carries out each of the following functions:
 ■ energy transduction
 ■ energy storage
 ■ information storage
 ■ transport of oxygen
 ■ carrying messages
 ■ protection against disease-causing organisms

5 Explain what makes carbon suitable as the 'element of life'.

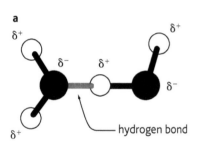

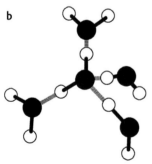

Figure 1.2.1 a *Two water molecules with a hydrogen bond between them.* **b** *Each water molecule may have hydrogen bonds with up to four other water molecules.*

∞ Link

Hydrogen bonds are important in cellulose; this will be dealt with on page 9. They are also important in DNA (page 63) and in stabilising proteins (page 15).

The giver of life

Water forms approximately 70 per cent of the bodies of animals, including humans, and makes up about 90 per cent of plants. All cells are surrounded by water and many organisms live in it. Life evolved in water and all organisms are dependent on it. At the temperatures we have on Earth water (H_2O) should be a gas like hydrogen sulphide (H_2S) and not a liquid. The reason it is not probably explains why life exists on Earth and not elsewhere in the solar system (as far as we know).

There are **hydrogen bonds** between water molecules. These exist because the oxygen atom has a greater attraction for the electrons in the covalent bond with hydrogen. This makes the oxygen slightly electronegative and the hydrogen slightly electropositive so that a water molecule is **dipolar**. The partial charges are indicated by the symbol δ (delta) with δ⁻ on the oxygen and δ⁺ on the hydrogen. A hydrogen bond forms between the slightly negative charge and the slightly positive charge. Each hydrogen bond is very weak (about one tenth the strength of a covalent bond) and easily broken. In bodies of water, hydrogen bonds break and reform all the time.

Water molecules are 'sticky' thanks to the hydrogen bonds between them. This **cohesion** between water molecules gives water important properties that are summarised in the table opposite.

☑ Study focus

You may be asked to discuss how the structure and properties of water explain the roles of water in living organisms and also as an environment. Remember that hydrogen bonding is key to these explanations.

Summary questions

1 Define the terms solvent; solute; soluble; insoluble; dipolar; cohesion.

2 Explain why hydrogen bonds form between water molecules.

3 Find the names of seven different biological molecules that dissolve in water and seven that do not.

4 Explain the importance of water as a component of cells, a transport medium and a coolant.

5 Discuss the advantages and disadvantages of water as an environment for organisms.

6 Discuss why the presence of liquid water on Earth is so important for life as we know it.

7 Find out what happened to water on the planet Venus.

Property	Explanation	Roles of water in organisms	Roles of water as an environment
good solvent for charged substances that dissolve readily in water; uncharged substances also dissolve, but less readily	polar molecules (e.g. glucose) and ions (e.g. Na^+, Cl^-) are charged and are attracted to the weak charges on water molecules	solvent within cells solvent in transport media, e.g. blood plasma; lymph; phloem and xylem saps	solvent for nutrients and gases (oxygen and carbon dioxide) carbon dioxide is much more soluble than oxygen
high specific heat capacity	4.2 J are necessary to increase the temperature of 1 g of water by 1 °C; this thermal energy breaks hydrogen bonds between water molecules	specific heat capacity of water is higher than that of other common substances this limits the fluctuations in the temperature of organisms and the environment of those that live in water	
high latent heat of vaporisation	much thermal energy is needed to cause water to change to water vapour	loss of water for cooling (e.g. in transpiration and sweating) is efficient as a lot of thermal energy is needed to evaporate small quantities of water	water in shallow aquatic habitats (e.g. ponds, rock pools) does not evaporate too quickly
high latent heat of fusion	much thermal energy is needed to change ice to water; much is transferred from water when it forms ice	water in cells tends to stay as a liquid, so cell membranes are not damaged by ice crystals	
reactive	water splits to form hydrogen ions (H^+) and hydroxyl ions (OH^-)	raw material for photosynthesis provides hydrogen ions and electrons for photosynthesis and respiration used in hydrolysis reactions, e.g. digestion	
density	ice is less dense than water	ice that forms in cells breaks cell membranes and kills cells; organisms at risk of freezing make 'anti-freeze' compounds that lower freezing point of cytoplasm	provides buoyancy for aquatic organisms so do not need highly developed skeleton ice floats on water – acts as an insulation for aquatic organisms beneath
incompressible	water cannot be compressed into a smaller volume	hydrostatic skeleton in animals, e.g. sea anemones, worms turgidity in plant cells, which provides support	
high cohesion	hydrogen bonds hold water molecules together	supports columns of water in xylem	gives surface tension – some organisms live on surface of water

1.3 Carbohydrates – sugars

Learning outcomes

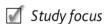

On completion of this section, you should be able to:

- describe the structure of the ring form of glucose
- state the difference between α and β glucose
- describe the structure of sucrose
- explain the relationship between the structure and function of glucose and of sucrose.

✓ Study focus

Simple sugars are reducing agents as they donate electrons from their aldehyde and ketone groups. They are known as **reducing sugars**.

∞ Link

In solution, glucose is in transition between the straight chain form and the ring form. In the straight chain form an aldehyde group is exposed giving the molecule its ability to act as a reducing agent. Fructose has a ketone group and is also a reducing sugar. For more information on this see page 18.

∞ Link

Pentoses are component molecules of nucleotides and nucleic acids (DNA and RNA) and these will be looked at in more detail on page 62. Trioses are important in the metabolism of organisms in respiration and photosynthesis, for example.

Carbohydrates are organic compounds that have the following properties, they:

- contain elements C, H and O
- have the general formula $C_x(H_2O)_y$
- include monosaccharides (e.g. glucose), disaccharides (e.g. sucrose) and polysaccharides (e.g. starch, glycogen and cellulose).

Simple sugars

Simple sugars are **monosaccharides,** which have the formula $C_x(H_2O)_y$ in which x is three, four, five, six or seven. Glucose with the formula $C_6H_{12}O_6$ is an example. Glucose molecules exist as straight chains and as rings. Most often glucose is in the ring form as shown here. There are two ring forms dependent on the position of the –H and –OH groups about the carbon atom at position 1. The two forms are α (alpha) with the –OH below the ring and β (beta) with the –OH above the ring. These two forms of glucose are polymerised to form macromolecules with very different properties and roles (see page 8).

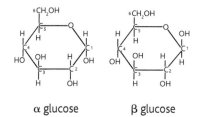

α glucose   β glucose

Figure 1.3.1 α and β glucose. The numbers refer to the carbon atoms

Glucose has six carbon atoms and so is an example of a **hexose** sugar; other hexoses are fructose and galactose. Glucose is a polar molecule and is therefore water soluble. It is the form of carbohydrate that is transported in the blood of animals. It is readily taken up by cells and metabolised:

- to provide a source of energy in respiration
- polymerised to form a **polysaccharide** for energy storage
- used as a raw material to make other compounds, e.g. **disaccharides**.

Other common monosaccharides that you will come across are pentoses that have five carbon atoms and trioses that have three.

Complex sugars

Monosaccharides are joined together to form complex sugars known as disaccharides. The diagram shows how this happens in plant cells that make sucrose from glucose and fructose. In the reaction, a molecule of water is removed so that an oxygen 'bridge' forms between C1 of the glucose molecule and C2 of the fructose molecule. The bond that forms is known as a **glycosidic bond** and the type of reaction is a **condensation reaction** because water is formed. The glycosidic bond is a type of covalent bond and is therefore very strong.

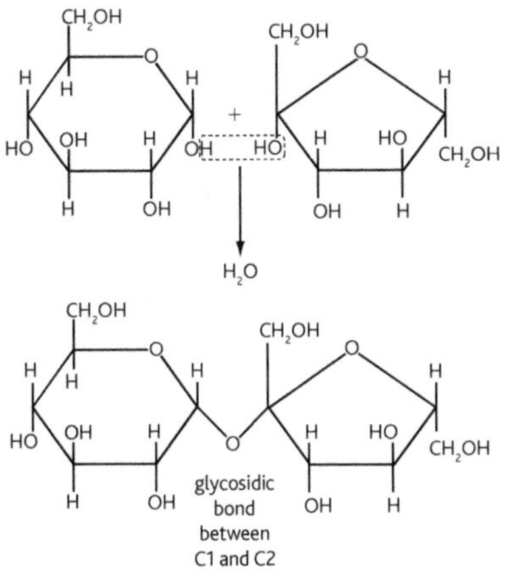

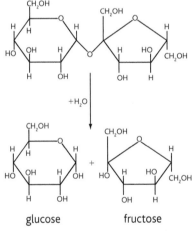

Figure 1.3.2 Formation of the disaccharide, sucrose

> ☑ *Study focus*
>
> When sucrose is boiled with Benedict's solution nothing happens. If sucrose is hydrolysed by reacting with hydrochloric acid or the enzyme sucrase then the glucose and fructose released react with Benedict's solution to give a colour change. See page 18 for more details.

In fact, sucrose is never formed in the way shown in Figure 1.3.2. The reaction is much more complex, but its formation does involve the elimination of water as shown. Sucrose is formed by plants for transport in the phloem. Sucrose is polar and water soluble, but not as reactive as glucose and fructose as the aldehyde and ketone groups form the glycosidic bond and are not available to react. This lack of reactivity makes sucrose a **non-reducing sugar** (see page 18). It may be an advantage to have a less-reactive sugar for transport in plants as the transport is slower than that of glucose in animals.

glucose   fructose

Figure 1.3.3 When the glycosidic bond in sucrose is broken by the addition of water, two monosaccharides are formed

The type of reaction in which water is added to break a glycosidic bond is a **hydrolysis reaction**.

Summary questions

1 Define the following terms: *carbohydrate; monosaccharide; hexose; disaccharide; glycosidic bond; condensation; hydrolysis reactions*.

2 Use simple diagrams to show the difference between α and β glucose.

3 Make a table to compare the structure and functions of glucose and sucrose.

4 Make simple diagrams to show the formation and breakage of a glycosidic bond between two hexoses.

5 Make a list of the features of carbohydrates.

6 Suggest why glucose is transported in animals, but sucrose is transported in plants.

Learning outcomes

On completion of this section, you should be able to:

- describe the structures of the polysaccharides: starch, glycogen and cellulose

- explain the relationship between the structure and function of the polysaccharides.

Did you know?

Cellulose is the most common organic compound on Earth. The bacteria, fungi and termites that recycle it as carbon dioxide play an important role in the biosphere.

Glucose and other monosaccharides are used as monomers to make polymers known as **polysaccharides**, which are used for energy storage and for making cellulose for cell walls. Polysaccharides are made from many monomers. Figure 1.4.1 shows a glucose monomer added to the end of an existing chain with the formation of a glycosidic bond.

addition of glucose to a growing end of amylose

1,4 glycosidic bond between C1 and C4

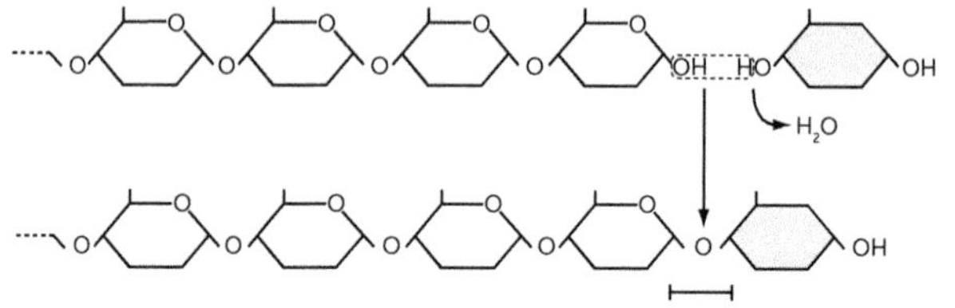

Figure 1.4.1 *The formation of a glycosidic bond between the end of a polysaccharide and a glucose monomer*

Glycosidic bonds that form between C1 at the end of the growing chain and C4 of the glucose monomer that is being added are known as 1,4 glycosidic bonds. If all the glucose monomers are added in this way an unbranched chain is formed. A branching point is formed by adding a glucose monomer to carbon 6 on a growing chain. The type of glycosidic bond that forms to make these branching points is a 1,6 glycosidic bond. From the first monomer added another chain can form with more 1,4 glycosidic bonds joining glucose monomers together. There are three energy storage polysaccharides:

- **amylose**
- **amylopectin**
- **glycogen**.

Amylose and amylopectin are forms of starch. Glycogen is very similar to amylopectin but has more 1,6 glycosidic bonds than amylopectin and therefore has many more branches. Glycogen is sometimes called 'animal starch'.

The three energy storage polysaccharides are made from α glucose monomers.

Cellulose is a long chain molecule made from β glucose monomers. It is not used for energy storage, but for making the cell walls of plants.

Polysaccharide	Monomer	Glycosidic bonds	Structure	Role
starch amylose	α glucose	1,4	unbranched chain – right-handed helix	energy storage in plants
amylopectin	α glucose	1,4 and 1,6	branched chain not a helix	
glycogen	α glucose	1,4 and 1,6	branched chain, more branched than amylopectin	energy storage in animals, fungi and some bacteria
cellulose	β glucose	1,4	unbranched straight chain	grouped into bundles within plant cell walls

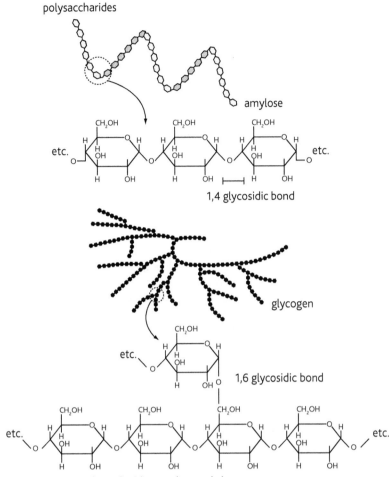

Figure 1.4.2 *Two polysaccharides: amylose and glycogen*

Study focus

To help you understand the similarities and differences between the biological polymers, make some simple diagrams of amylose, amylopectin and glycogen and annotate them with points about their structure and function. You can do the same for proteins and nucleic acids.

Structure and function

The storage polysaccharides are insoluble in water, unreactive and compact. Amylose and amylopectin are stored in plant cells as starch grains and glycogen is stored in animal cells as smaller granules. As amylopectin and glycogen have branches they have many places where glucose can be added or removed as required by a cell.

Alternate β glucose molecules are arranged at 180° to each other as they are added to a growing cellulose molecule. This gives a straight chain, not a helix, with many projecting –OH groups on both sides of the chain to form hydrogen bonds with adjacent cellulose molecules. A bundle of cellulose molecules bonded together by hydrogen bonds forms a microfibril that is very strong. Microfibrils are arranged in plant cell walls in a criss-cross pattern to provide even more strength.

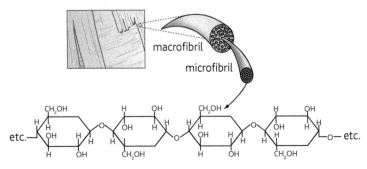

Summary questions

1 Explain why starch and glycogen make suitable molecules for energy storage and cellulose is suitable for cell walls of plant cells.

2 Draw a diagram to show the formation of a 1,6 glycosidic bond to make a new branching point in glycogen.

3 Find out the names of the following: **i** a polymer of fructose; **ii** a polysaccharide made of two or more monomers; **iii** components of plant cell walls other than cellulose; **iv** cell wall components of bacteria and fungi.

4 Explain how glycogen and cellulose are suited to their functions.

Figure 1.4.3 *Close packing of cellulose molecules in a microfibril and the pattern of microfibrils in a cell wall*

On completion of this section, you should be able to:

- describe the molecular structure of a triglyceride

- explain how triglycerides are good sources of energy

- describe the molecular structure of a phospholipid

- explain why phospholipids are suitable as the main component of biological membranes.

Lipids are organic compounds composed of the same elements as carbohydrates, but they have a much higher ratio of hydrogen to oxygen. In addition they do not have the same structure as polysaccharides with repeated monomers. The two groups of lipids described here, **triglycerides** and **phospholipids**, have the same basic structure with glycerol and three attached units. These units are fatty acids. Phospholipids also have phosphate and nitrogen-containing groups attached. The carboxylic acid group of fatty acids reacts with the –OH groups of glycerol to form ester bonds.

Triglycerides

Each triglyceride molecule consists of glycerol and three fatty acids. There are different fatty acids, some are **saturated fatty acids**, such as **A** in Figure 1.5.1, as they have the full complement of hydrogens attached to the carbon chain. Saturated fatty acids do not have any double bonds between carbon atoms in the chain. Some fatty acids are unsaturated, such as **B** in Figure 1.5.1, with at least one double bond between carbon atoms along the carbon chain and therefore fewer hydrogens. Some triglycerides have three identical fatty acids; others have a mixture of different fatty acids.

As lipids have many –CH groups rather than –OH groups they are not polar and are insoluble in water. Triglycerides make excellent long-term energy storage molecules. They are stored in special fat tissue (adipose tissue) in animals and as droplets of oil in plants. Seeds are especially rich in oils. Fats and oils are more efficient for energy storage than carbohydrates as they are highly reduced molecules because of all those hydrogens. When oxidised, during respiration, much more energy is released than from the same mass of carbohydrate or protein.

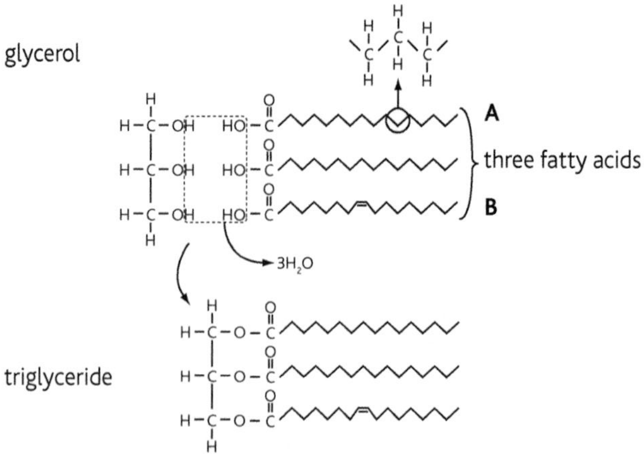

Figure 1.5.1 Glycerol and three fatty acids combine to form a triglyceride

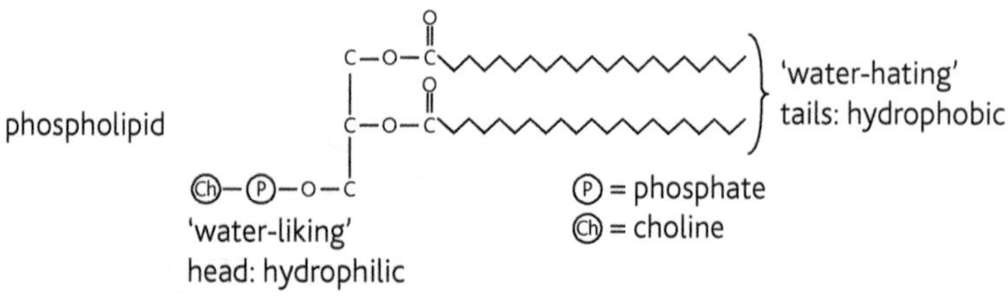

Figure 1.5.2 A phospholipid is composed of glycerol, two fatty acids and a phosphate-containing part

Phospholipids

Each phospholipid molecule consists of glycerol and two fatty acids. Attached to one of the –OH groups of glycerol is a phosphate group and often to that a nitrogen-containing group, such as choline. This phosphate 'head' is water soluble, whereas the two fatty acids are not soluble in water. This makes the molecule have a hydrophilic ('water-liking') region and a hydrophobic ('water-hating') region. If in contact with water a single layer of phospholipids will either form a layer on top of the water or will form tiny spheres with the hydrophilic heads attracted to water and the hydrophobic tails in the centre.

Two layers of phospholipids will form a bilayer with a central hydrophobic region and two hydrophilic regions in contact with water. This is the basis of the phospholipid bilayer that is the structure of biological membranes, which will be covered on page 36.

Fat in the diet and obesity

Fat in the human diet is a good thing. There are two fatty acids that we need which we cannot make from anything else and they are known as essential fatty acids. Some vitamins, such as vitamin D, are fat soluble so we need to have fat in the diet otherwise we would be deficient in these vitamins and suffer ill health. Fat is also a good provider of energy and is stored for protecting organs, providing energy during periods when there is little or no food and for thermal insulation. Humans are good at storing fat and can convert excess carbohydrate and protein into fat when food is scarce. During periods when food is scarce people are in negative energy balance and use more energy than they can obtain in their diet. People who have a positive energy balance store the excess energy as fat and may become overweight or even obese. Many people do not experience food shortages; in fact they have more than enough food to eat. As a result in many countries there is an epidemic of obesity with over 30 per cent of the population being classified as obese or very obese.

Did you know?

There are a variety of ways to classify people as obese including: body mass 20% or more above the recommended mass for height; body mass index (BMI) greater than 30.

Summary questions

1   Define the following terms: *fatty acid; glycerol; ester bond; saturated fatty acid; unsaturated fatty acid; hydrophilic; hydrophobic; monolayer; bilayer; obesity*.
2   Make a table to compare the structure and functions of triglycerides and phospholipids. (Remember to include what they have in common as well as the differences between them.)
3   Explain why triglycerides are efficient sources of energy and good for long-term storage.
4   Explain the importance of phospholipids in organisms.
5   Find: **i** the names of the two essential fatty acids in the human diet; **ii** the fat soluble vitamins; **iii** the risks to health of obesity.
6   The body mass index (BMI) is calculated using the following formula:

$$BMI = \frac{body\ mass\ in\ kg}{(height\ in\ metres)^2}$$

   a   Calculate the BMI for the following people:
      A 1.65 m, 45 kg; B 1.63 m, 64 kg; C 1.65 m, 75 kg; D 1.45 m, 75 kg.
   b   Use the table to identify the appropriate categories for each person.
   c   What advice would you give these four people?

BMI	Category
below 20	underweight
20–35	acceptable
25–30	overweight
over 30	obese
over 40	very obese

1.6 Proteins (1)

Learning outcomes

On completion of this section, you should be able to:

- describe the generalised structure of amino acids

- state that there are 20 different amino acids used to make proteins

- explain how amino acids differ from one another

- describe the formation and breakage of a peptide bond.

Amino acids

Proteins are macromolecules made from one or more polypeptides. A polypeptide consists of an unbranched chain of amino acids. There are 20 different types of amino acid that cells use to make proteins. However, they all share the same molecular structure as shown in Figure 1.6.1.

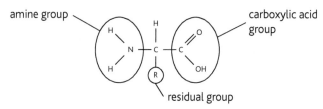

Figure 1.6.1 *A generalised amino acid molecule*

All amino acids have a central carbon atom, an amine group and a carboxylic acid group. Attached to the central atom is a hydrogen atom and a group that is specific to the type of amino acid. If it is another hydrogen atom then the amino acid is glycine, which is the smallest. If the group is $-CH_3$ then the amino acid is alanine (see Figure 1.6.2). In the generalised amino acid this group is known as R for residual.

Amino acid	Three-letter abbreviation	R group	Comments
glycine	gly	$-H$	smallest amino acid; many found in collagen to allow for close packing
methionine	met	$-CH_2-CH_2-S-CH_2$	first amino acid of every primary sequence when a polypeptide is produced on a ribosome (see page 71)
alanine	ala	$-CH_3$	non-polar R group
phenylalanine	phe	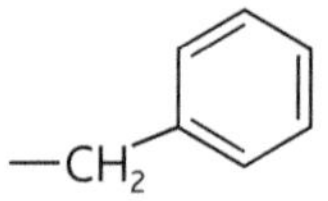 $-CH_2$ ⬡	non-polar R group, which has a ring structure formed of carbon atoms
cysteine	cys	$-CH_2-SH$	sulphur-containing R group; forms covalent disulphide bonds between parts of a polypeptide or between polypeptides

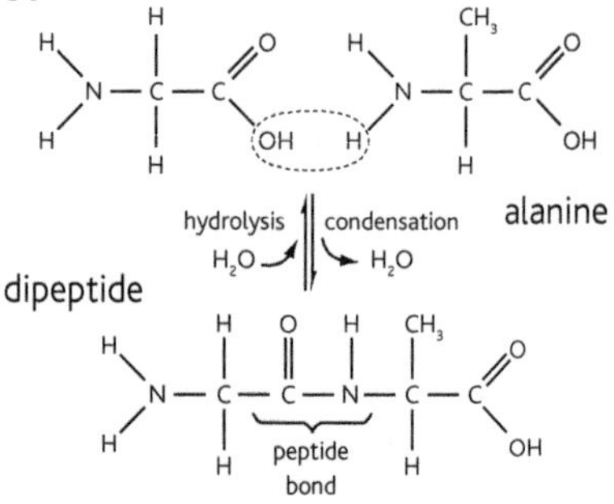

Figure 1.6.2 *The formation and breakage of a peptide bond*

When two amino acids are joined together a peptide bond forms linking the C of the carboxylic group of one amino acid with the N of the amine group of the other (see Figure 1.6.2). The addition of another amino acid forms a tripeptide. A polypeptide consists of ten or more amino acids.

antidiuretic hormone	Cys	Tyr	Phe	Gln	Asn	Cys	Pro	Arg	Gly	–NH₂
oxytocin	Cys	Tyr	Ile	Gln	Asn	Cys	Pro	Leu	Gly	–NH₂

Figure 1.6.3 *Two nonapeptides (with nine amino acids) that are biologically active.*
(–NH$_2$ represents the N terminal of the peptide; the opposite end is the C terminal.)

R groups

The properties of polypeptides and proteins are dependent on the R groups that project from the central chain of: –C–N–C–C–N–C–C–N–.

As you can see from the table some amino acids have polar groups so attract water molecules, but others are non-polar and do not. This means that polypeptides can have a sequence of amino acids with non-polar R groups to form a hydrophobic region that can pass through the central part of a phospholipid bilayer.

Some R groups ionise to form negatively or positively charged groups. These attract to each other to form ionic bonds.

The position of cysteine in polypeptides is important as two adjacent –SH groups can react together to form a disulphide bond. These act to form firm attachments to link different parts of a polypeptide together or to link different polypeptides together.

The position of amino acids in a polypeptide is not random. The sequence of amino acids is determined by genes as you will discover on page 66. If the sequence is changed then the structure will be different and maybe the polypeptide will not function or will function differently. Some proteins have special areas, which are like 'pockets' in the molecule. Other molecules fit into these 'pockets':

- substrates fit into the 'pockets' in enzymes
- signalling molecules, such as hormones, fit into binding sites in receptor molecules
- antigens fit into binding sites in antibody molecules.

Any change to the amino acids in these 'pockets' changes the shape and the charge distribution. This means that the molecules, such as substrates, messenger molecules and antigens will not fit and the proteins will be non-functional.

Did you know?

The artificial sweetener aspartame is 200 times sweeter than sugar. It is a dipeptide of two amino acids.

∞ Link

Proteins that pass through membranes have a hydrophobic region, for more information on this see page 36.

∞ Link

The 'pockets' in enzymes are active sites, for more information on them refer to page 49.

☑ Study focus

You are not required to remember the different types of amino acids, but if you can it helps to explain the properties of proteins as highlighted in the next section.

Summary questions

1 Define the following:
   - amino acid
   - peptide bond
   - dipeptide
   - tripeptide
   - polypeptide.

2 Make an annotated drawing of the amino acid cysteine. Explain the importance of this amino acid in proteins.

3 Make a diagram to show how a tripeptide is formed from a dipeptide and an amino acid.

4 Give three reasons why the sequence of amino acids in a protein is important.

5 Find the names of: **i** the two amino acids that form aspartame; **ii** some amino acids that are not used in the production of proteins and give their functions.

6 Write short profiles of the nonapeptides antidiuretic hormone and oxytocin.

Learning outcomes

On completion of this section, you should be able to:

- describe the four levels of organisation in a protein
- describe the bonds that hold proteins in shape.

Levels of organisation

A polypeptide is formed by joining ten or more amino acids together. Each polypeptide has a definite number of amino acids. They are formed as straight chains which spontaneously form into specific shapes.

The table summarises the levels of organisation.

Level of organisation	Description	Comments
primary structure	■ sequence of amino acids ■ position of disulphide bonds	determined by the gene that codes for the protein position of cysteines in sequence determines where these will form
secondary structure	α-helix β-pleated sheet	polypeptide forms a right handed helix polypeptide folds back and forth to form a flat sheet
tertiary structure	further folding of polypeptide to give complex 3D shape	some polypeptides have both α-helices and β-pleated sheets
quaternary structure	two or more polypeptides associate together to form a protein	the polypeptides can be identical or different

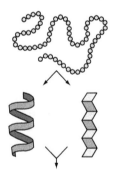

Primary structure – sequence of amino acids and position of disulphide bonds

Secondary structure – α-helix and β-pleated sheet (the latter are drawn as wide arrows in ribbon models of proteins)

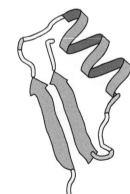

Tertiary structure – further folding to give a complex structure with α-helices and β-pleated sheets, as well as regions without distinct secondary structure

Figure 1.7.1 *The three levels of organisation of a polypeptide*

Did you know?

Some proteins have alpha helices, but no beta-pleated sheets. An example is haemoglobin, which is covered on page 16. Some proteins are composed almost entirely of beta-pleated sheets – for example, fibroin that is a major component of silk spun by silkworms and spiders.

Bonds that stabilise proteins

There are four bonds that stabilise polypeptides:

- hydrogen bonds form between polar groups, such as the dipolar –NH and –CO
- ionic bonds form between ionised amine and carboxylic acid groups
- hydrophobic interactions between non-polar side chains
- disulphide bonds between the S-containing R groups of cysteines.

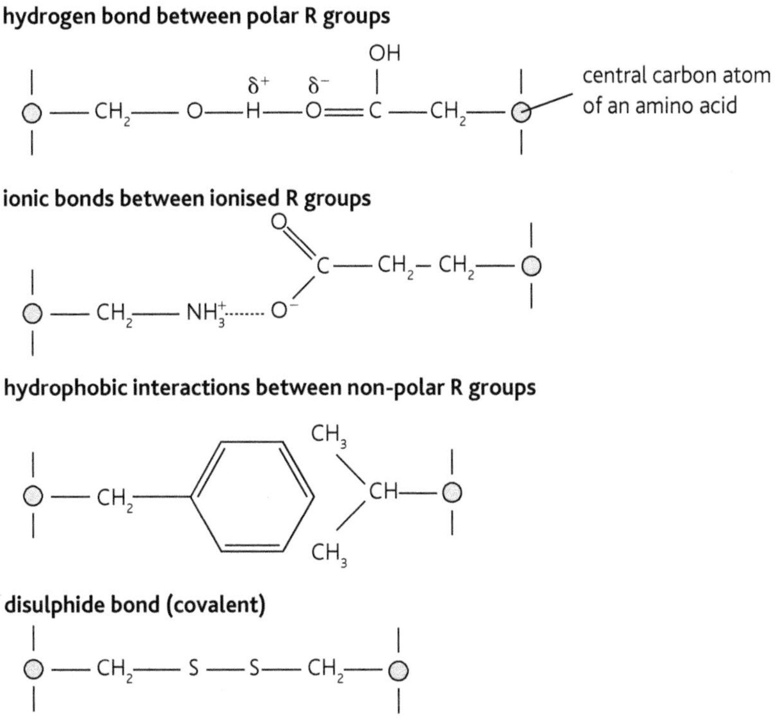

hydrogen bond between polar R groups

central carbon atom of an amino acid

ionic bonds between ionised R groups

hydrophobic interactions between non-polar R groups

disulphide bond (covalent)

Figure 1.7.2 The bonds that stabilise polypeptides

After proteins are produced inside cells by the assembly of amino acids, they are further processed. There are a variety of ways in which this might happen:

- The addition of a prosthetic group that is part of a protein not composed of amino acids; for example, haem that is the prosthetic group of haemoglobin and catalase.
- Polypeptides are assembled to give a protein its quaternary structure; haemoglobin and catalase both have four polypeptides each with a haem group.
- Chains of sugar molecules may be added to the polypeptide to form a glycoprotein.
- Polypeptides may be cut into two or more pieces and joined together, as happens in the formation of insulin.

Globular proteins are soluble in water with hydrophilic R groups on the surface forming hydrogen bonds with water molecules. Internally there are hydrophobic R groups that exclude water. Fibrous proteins are insoluble in water.

 Link

For more information about hydrogen bonds, see page 4.

Summary questions

1 Find a ribbon diagram of the protein lysozyme. Annotate it to show the regions of secondary structure and explain why it has tertiary structure but not quaternary structure. State the function of lysozyme.

2 Find out how many polypeptides there are in the following proteins: α amylase; insulin; glucagon; catalase; carbonic anhydrase; myoglobin. State the functions of each of these proteins.

3 Explain the difference between the primary structure of a protein and its secondary structure.

✓ *Study focus*

Diagrams of haemoglobin sometimes show the molecule within a circle. This makes it look as if one haemoglobin molecule fills a red blood cell. It is estimated that each red blood cell contains about 280 million haemoglobin molecules, not one!

Did you know?

Haem is an example of a prosthetic group – a part of a protein molecule that is not made of amino acids. Proteins like this, containing parts that are not made of amino acids, are called conjugated proteins.

✓ *Study focus*

Remember that any protein that has more than one polypeptide has quaternary structure.

Globular and fibrous proteins

Globular proteins are soluble in water and folded into complex 3-D shapes. **Fibrous proteins** are insoluble in water and have simple shapes, such as a helix. **Haemoglobin** is a globular protein found inside red blood cells; **collagen** is a fibrous protein that is a major component of the material between cells in structures such as tendons, ligaments, muscles and bone. Both are formed from more than one polypeptide.

Haemoglobin

Each molecule of haemoglobin is composed of four polypeptides – two α globins and two β globins. In the centre of each polypeptide is a haem group. A haem group is not made of amino acids; chemically it is very different with an atom of ferrous iron (Fe^{2+}) at its centre. Each of the haem groups forms a temporary bond with an oxygen molecule. As there are four haem groups this means that each haemoglobin molecule can carry four molecules of oxygen. The addition of the first molecule of oxygen changes the shape of haemoglobin making it easier to accept the second oxygen molecule. This makes it easier to accept the third and in turn this makes it easier to accept the fourth. This is because the molecule changes shape opening out from a 'tense' form to a more 'relaxed' form exposing the haem groups to accept oxygen.

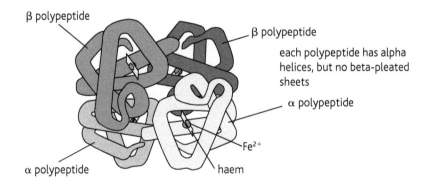

β polypeptide

β polypeptide

each polypeptide has alpha helices, but no beta-pleated sheets

α polypeptide

Fe^{2+}

α polypeptide

haem

Figure 1.8.1 *Haemoglobin has quaternary structure*

Collagen

Collagen is an extracellular protein that provides toughness to skin, bone, cartilage, tendons, ligaments and muscles. Collagen has several levels of organisation that are not quite the same as those of haemoglobin.

A molecule of collagen is made of three identical polypeptides that form left-handed helices. These are wound around each other to form a triple helix. The polypeptides are long, consisting of about 1000 amino acids with glycine as every third amino acid. Glycine has the smallest R group (–H) so it does not take up much space. This means the helices can be wound tightly together and form many hydrogen bonds between them. The triple helix is not folded again as in the tertiary structure of haemoglobin. The triple helices are joined together by covalent bonds to form a network of fibres. The ends of the triple helices do not coincide within the fibres so there are no lines of weakness where the fibres may break. This arrangement makes collagen suitable for structures such as tendons that have high tensile strength and resist pulling forces.

∞ *Link*

Cellulose is a polysaccharide and collagen is a protein, but both have similar structural features and functions. Try comparing their structure and function by completing Summary question 8 below.

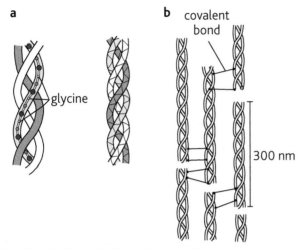

Figure 1.8.2 a *The triple helix of collagen;* **b** *triple helices are joined together to form collagen fibres*

Figure 1.8.3 *Collagen fibres from human skin. Notice the characteristic banding pattern that is visible at this magnification in an electron microscope (the fibres are 300 nm wide).*

Summary questions

1 Explain why haemoglobin is an example of the following: globular protein; conjugated protein; protein with quaternary structure.

2 Explain how the structure of haemoglobin is related to its transport functions.

3 Explain why collagen is an example of a fibrous protein.

4 Collagen and fibroin (see page 14) are both fibrous proteins. State how their structures differ.

5 Explain how the structure of collagen is related to its structural functions.

6 Draw a table to compare the structure, distribution and functions of haemoglobin and collagen.

7 Find some computer-generated models of haemoglobin and collagen and annotate them to show their structure.

8 Draw a table to compare the structure, location and function of cellulose and collagen.

Learning outcomes

On completion of this section, you should be able to:

- describe the biochemical tests for starch; reducing sugars; non-reducing sugars; proteins; lipids
- state the positive and negative results for these tests.

Test for starch

The reagent for the starch test is the iodine in potassium iodide solution (known simply as iodine solution).

1 The substance to be tested may be a solid or a liquid.
2 If solid, place a sample of the substance on a white tile or in a Petri dish.
3 If liquid place into a test-tube.
4 Use a dropping pipette to add iodine solution.

Colour change with iodine solution	Result	Explanation
yellow-orange to blue-black or blue	positive starch present	iodine binds to the centre of the helix of amylose to form starch-iodine complex which has a blue-black colour
no change; iodine solution remains yellow-orange	negative no starch present	no starch for iodine to bind to

✓ *Study focus*

Remember to call the reagent for the starch test 'iodine solution' not 'iodine'.

Test for reducing sugars

The reagent for the reducing sugar test is Benedict's solution – an alkaline solution of copper sulphate.

1 If the substance to be tested is solid, make a solution with water.
2 Put about $1\,cm^3$ of the test solution into a test-tube and add an equal volume of Benedict's solution.
3 Boil in a water bath (do not heat directly with a Bunsen burner).
4 Watch carefully for colour changes.

(The test-tube may be put into a water bath at about $80\,°C$ rather than placing into a water bath with boiling water.)

Colour change on boiling with Benedict's solution	Result	Explanation
blue to green/yellow/orange/red with a precipitate	positive reducing sugars present (not necessarily glucose)	sugar reduces copper(II) ions (Cu^{2+}) in Benedict's solution to copper(I) ions (Cu^+) to form a precipitate of copper(I) oxide
no change; Benedict's solution remains blue	negative	no reducing sugars present to react with copper(II) ions

Test for non-reducing sugars

If a test substance gives a negative result with the reducing sugar test it may contain non-reducing sugars. The only common non-reducing sugar is sucrose (see page 7).

1 Divide the test solution into two equal parts, A and B.
2 Test A with Benedict's solution as above.

3 Add a few drops of dilute hydrochloric acid to B and boil for a few minutes.
4 Cool the test-tube and add dilute sodium hydroxide solution or solid sodium hydrogen carbonate (beware, the latter will fizz).
5 When neutralised, test with Benedict's solution as above.

Colour change on boiling with Benedict's solution	Result	Explanation
A – no change B – blue to green/yellow/ orange/red	negative for reducing sugar, positive for non-reducing sugar	hydrochloric acid acts as a catalyst to hydrolyse sucrose to fructose and glucose that are both reducing sugars
A – no change B – no change	negative for both reducing and non-reducing sugars	no reducing sugars present even after using hydrochloric acid to hydrolyse any non-reducing sugars

Test for proteins

The reagent for the protein test is biuret solution (a solution of copper sulphate and sodium hydroxide).

1 If the substance to be tested is solid, make a solution with water.
2 Place 1 cm³ of the test solution in a test-tube.
3 Add the same volume of biuret solution and mix by shaking the tube from side to side.

Colour change with biuret solution	Result	Explanation
blue to violet/purple/ lilac	positive	a coloured complex forms where there are peptide bonds
no change to the blue colour	negative	no peptide bonds present

Test for lipids

Lipids are insoluble in water, but they are soluble in organic solvents such as ethanol. This test makes use of this fact.

1 Crush any solid material to be tested in a pestle and mortar and add some ethanol.
2 If the test substance is a liquid just add some ethanol and shake to dissolve.
3 Pour off the ethanol, which may have dissolved some lipids, into a test-tube of water (do not mix).

Change when adding ethanol to water	Result	Explanation
white cloudiness – an emulsion	positive	the ethanol dissolves the lipid; on addition to water the lipid is dispersed throughout the water as tiny particles – an emulsion
no change	negative	no lipid present to be dispersed

☑ *Study focus*

You should carry out these tests for yourself in the lab. Make sure you record and learn all the details as you may be tested on them in the examination. It is easy to forget practical procedures.

☑ *Study focus*

What if samples contain *both* reducing and non-reducing sugars? What would be the results for samples A and B? Attempt Summary question 4 and see if you are right.

Summary questions

1 Make a summary table to show these biochemical tests. Use the following column headings: test biochemical (starch, etc.); reagent; method; positive result; negative result.

2 Make a flow chart to show how to test a sample of plant storage tissue for reducing and non-reducing sugars.

3 You have a solution that is a mixture of glucose and fructose. Explain why the Benedict's test cannot be used to confirm this.

4 Some plant material contains both glucose and sucrose. Explain how you can use the Benedict's test to confirm this.

1.10 Quantitative biochemical tests

Learning outcomes

On completion of this section, you should be able to:

- explain how to carry out a semi-quantitative test for reducing sugars
- describe how to carry out a quantitative test for starch.

The tests in Section 1.9 are all qualitative. They tell you that the biochemical is present but not how much is present. The tests described here attempt to make the results quantitative to varying degrees.

Semi-quantitative Benedict's test

The final colour change with Benedict's test gives an idea of how much reducing sugar is present in a test sample. For example, if the final colour is green then the concentration of the reducing sugar is very low; if red it is much higher. One way to improve this estimate is to make up a series of colour standards using a glucose solution and Benedict's solution.

1 Take $20\,cm^3$ of a stock glucose solution of known concentration, e.g. $100\,g\,dm^{-3}$.
2 Make a series of dilutions from this stock solution, e.g. 50.0, 20.0, 10.0, 5.0, 1.0, 0.5, $0.1\,g\,dm^{-3}$.
3 Place equal volumes of the dilutions into labelled test-tubes.
4 Carry out the Benedict's test as in Section 1.9 (using equal volumes and heating or boiling all the test-tubes for the same length of time).
5 Cool the test-tubes and keep them (maybe take photographs for a permanent record).

You now have a set of colour standards.

Study focus

It is important to carry out the procedure in *exactly* the same way as when making the colour standards so that you know the results are valid. If you do them differently you cannot be sure about the accuracy of your estimates of the concentrations.

Concentration of glucose/$g\,dm^{-3}$	Result on testing with Benedict's solution
50.0	dark red
20.0	red
10.0	orange
5.0	yellow
1.0	green
0.5	light green
0.1	blue

6 Carry out the Benedict's test on a solution of the test substance using exactly the same procedure as when making the colour standards.
7 Cool the test-tube and place next to the colour standards to determine the concentration of reducing sugar; the answer may be a range, e.g. between 1.0 and $5.0\,g\,dm^{-3}$.

A test like this, which can give you an estimate of the concentration, is semi-quantitative.

This test can be improved by removing the precipitate, drying it and then weighing it on a balance. The masses recorded can be plotted on a graph against the concentration of glucose. The concentration of reducing sugar in any test substance can be determined by taking an intercept on the graph. You can try this for yourself in Question 4 on page 25.

Study focus

Summary question 1 gives you some results from testing some fruit juices for reducing sugar. Answer the question before continuing.

Quantitative test for starch

To make the iodine test for starch quantitative we can make a series of dilutions of a starch solution.

1 Take $20\,cm^3$ of a stock starch solution of known concentration, e.g. $100\,g\,dm^{-3}$.

2 Make a series of dilutions from this stock solution, e.g. 50.0, 10.0, 5.0, 1.0, 0.5, 0.1, 0.05 and $0.01\,g\,dm^{-3}$.

3 Place equal volumes of the dilutions into labelled test-tubes.

4 Add the same volume of iodine solution to each test-tube (remember it is not necessary to heat these as with Benedict's solution).

5 Place each test-tube into a colorimeter, which detects the optical density of the solutions (results are recorded as absorbance or percentage transmission).

6 Plot a graph of the colorimeter readings against the concentration of starch.

7 Follow exactly the same procedure with any test sample.

8 The concentration of starch in the test sample can be found by taking an intercept on the graph.

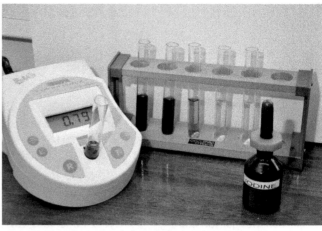

Figure 1.10.1 *Testing a solution of starch with a colorimeter*

∞ Link

You can read more about using a colorimeter to determine the concentration of starch in a solution on page 52.

Summary questions

1 A student tested some fruit juices with Benedict's solution with the following results:

Fruit juice	Colour after boiling with Benedict's solution
P	orange-red
Q	green
R	blue

Use the table on the opposite page to estimate the concentration of reducing sugars in the three fruit juices. Explain the limitations of this method of finding out the actual concentration in any test substance.

2 Explain why the Benedict's test cannot show whether or not fructose is in the fruit juices.

3 Make a table to show how to make the following dilutions from a $100\,g\,dm^{-3}$ starch solution:

   50.0, 10.0, 5.0, 1.0, 0.5, 0.1, 0.05 and $0.01\,g\,dm^{-3}$.

4 Explain the difference between qualitative and quantitative tests for reducing sugars.

5 Draw a flow chart to show the procedure that you would use to determine the decrease in concentration of starch in bananas as they ripen.

Learning outcomes

On completion of this section, you should be able to:

- recognise the multiple-choice questions that test knowledge and those that test understanding with knowledge
- list the different ways in which short answer questions are written
- plan answers to the longer answer questions.

☑ Study focus

Question 1 requires you to know about the levels of organisation of proteins. You do not have to work out anything from the question. A molecule of haemoglobin is made of four globins (polypeptides) so it has all four levels of organisation and **D** is the right answer.

☑ Study focus

Look at Section 1.4 to check your answer. This is an example of a multiple-choice question with a negative statement. Look out for these in the exam, as often they are the most difficult.

Your Unit 1 test consists of multiple-choice questions, short-answer questions and questions requiring longer answers.

In this section we will show you what these three types of questions are like based on topics in this chapter. In the next section there are more questions to test your knowledge and understanding of this chapter. There are also questions on the CD that relate to each chapter.

Multiple-choice questions (MCQs)

The MCQs will test your recall of knowledge and your ability to apply your understanding of the subject areas you have covered. The MCQs in this section are labelled with K (knowledge) and U of K (understanding of knowledge) to show you where they apply. They will NOT be labelled like this in the examination. In the examination you put your answer on a special answer sheet.

1   Haemoglobin is an example of a protein with:
    A   primary structure only
    B   primary and secondary structure only
    C   primary, secondary and tertiary structure only
    D   primary, secondary, tertiary and quaternary structure   (K)

2   Which of the following shows the correct bond associated with each of the biological molecules shown?

	Disaccharide	Polysaccharide	Protein	Triglyceride
A	ester	glycosidic	covalent	peptide
B	peptide	ester	glycosidic	covalent
C	glycosidic	glycosidic	peptide	ester
D	glycosidic	peptide	ester	glycosidic

(K)

Question 2 requires you to match up the macromolecules and the bonds that hold the sub-units together. It is a good idea to read each column in turn. Circle what you think is the right answer for the first column and then move on to the next. The right answer is **C**.

3   Starch and glycogen are polysaccharides. Which one of the following is a feature of starch but not of glycogen?
    A   Starch is made from α glucose
    B   Starch has an unbranched component
    C   Starch contains 1,6 glycosidic bonds
    D   Starch contains 1,4 glycosidic bonds   (U of K)

In each case, you have to work out whether the statement applies to glycogen or not. The only statement that does not apply is **B**.

Short answer questions (SAQs)

The questions in Paper 2 are answered in a different way and involve a variety of different responses:

- completing tables
- matching pairs (see Question **1** on page 24)
- completing sentences (Question **4**)
- writing one word answers (Question **5b**)
- writing sentences (Question **6a** and **b**)
- giving labels on a diagram or drawing
- describing a pattern or trend from a table or graph
- interpreting information in the form of text, table, graph or diagram.

4   Copy and complete the following passage by using the most suitable word(s) in each of the spaces provided.

Water is essential for life. It makes up most of the cytoplasm in cells and is the main component of body fluids, such as blood. Many different substances dissolve in water so making it an excellent … . The … bonds that form between water molecules are responsible for the … between those molecules which explains why water is able to travel to tops of very tall trees. They are also responsible for the … at the boundary between water and air. Water remains in the … state over a wide range of temperatures as it has a high … .

5   **a**   State what is meant by the term primary structure of a protein. Collagen and haemoglobin are two proteins.
   b   State TWO types of bonds, other than peptide bonds, that collagen and haemoglobin have in common.
   c   State THREE ways in which the structure of collagen differs from the structure of haemoglobin.

6   Triglycerides and phospholipids are found in animal and plant cells.
   a   Explain how the structure of triglycerides makes them more suitable for energy storage than carbohydrates, such as glycogen.
   b   Explain how the structure of phospholipids makes them suitable for cell membranes.

Longer answer question

Paper 2 also has some longer answer questions and each of these has a total of 15 marks.

7   Carbohydrates and proteins are biological molecules.
   a   Describe the structure of glucose and sucrose.
   b   Explain why glucose is soluble in water, but starch is not.

Proteins perform many roles inside cells; starch is an energy storage molecule.
   c   **i**   Explain how the structure of a protein differs from the structure of starch.
       ii   Explain why proteins perform multiple roles, but starch only performs one.

☑ *Study focus*

Question 5 shows you why it is important to learn the definitions of the terms given in the syllabus. When answering part **b** think about a table comparing the two proteins. For each difference you give make sure you write something about *both* collagen and haemoglobin, e.g. collagen has three polypeptide chains forming a triple helix, but haemoglobin has four polypeptides.

Notice that these questions ask about the relationship between structure and function. In both **a** and **b** make at least three points as there are three marks available. Describe the structure and then link it to the function. Refer to the glossary for appropriate technical terms.

☑ *Study focus*

Question 7 is not an essay question, but you will need to structure your answer carefully, deciding on a strategy to answer the question. You might want to give numbered points or a table if the question asks you to compare one or more items. It is always a good idea to plan your answer first, you can cross it out once you have written your answer in full.

1.12 Practice exam-style questions: Biochemistry

Try the following questions as examination practice.

1 Figure 1.12.1 shows eight biological molecules. Study them carefully and then answer the questions that follow.

In each question, choose one of the biological molecules and write down the letter it relates to. You may choose each molecule for more than one question and there may be one or more letters that you do not use at all.

a an amino acid that is a major constituent of collagen

b disaccharide found in the phloem

c molecule that is polymerised to form cellulose

d molecule with a peptide bond

e molecule that is hydrolysed to form fatty acids

f molecule with hydrophilic and hydrophobic regions

g an amino acid that forms disulphide bonds in proteins

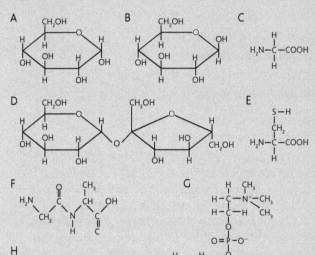

Figure 1.12.1

2 A student tested some plant and animal tissues to find out which biological molecules they contained. The results of the biochemical tests are shown in the table.

Tissues	Results of biochemical tests			
	Benedict's solution	Iodine solution	Biuret reagent	Ethanol and water
A	red precipitate	yellow-brown	lilac	white emulsion
B	blue	blue-black	blue	no emulsion
C	green	blue-black	lilac	no emulsion
D	yellow precipitate	yellow-brown	lilac	white emulsion
E	blue	yellow-brown	blue	white emulsion

a Use the information in the table to state which tissues:
  i contain starch
  ii did not contain any reducing sugars
  iii contain lipids and proteins.

b State which tissues are likely to be plant in origin and explain your answer.

c What conclusions can be made about the relative concentrations of reducing sugars in the five tissues? Explain your answer.

3 The table below includes statements about the roles of water:
- in living organisms
- as an environment for living organisms.

Copy and complete the table by indicating with a tick (✓) which one of the properties of water is responsible for each role.

You should put only one tick in each row.

Roles of water	Properties of water			
	High specific heat capacity	Strong cohesive forces between water molecules	High heat of vaporisation	Solvent for polar molecules and ions
transport medium in blood plasma and phloem				
surface for small insects to walk on				
major component of sweat used in heat loss				
movement of water in xylem				
prevents wide variations in body temperature				

4 A student carried out a quantitative test for reducing sugars by testing known concentrations of glucose with Benedict's solution. After the test, the reaction mixtures were filtered, using filter paper of known mass, to remove the precipitate. The filter paper with the precipitate was dried and then weighed. The results are in the table.

Concentration of glucose/$g\,dm^{-3}$	Initial mass of filter paper/g	Mass of filter paper and precipitate after drying/g	Difference in mass/g
0	0.84	1.16	0.32
20	0.83	1.17	0.34
40	0.81	1.17	0.36
60	0.82	1.27	0.45
80	0.84	1.24	0.40
100	0.83	1.27	0.44

a State which result is an anomaly.

b Describe the precautions that the student should take to make sure that valid and reliable results are collected.

c Suggest how the student's method can be used to find the reducing sugar concentration of a fruit juice.

d Explain why the student would not be able to state the glucose concentration of the fruit juices using this method.

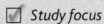 *Study focus*

Approach questions logically, for example if part **b** of this question has six marks allocated you should write at least six precautions.

5 a Describe THREE ways in which polysaccharides differ from polypeptides.

b Explain why the energy available in glycogen can be made available very quickly.

c Explain why cellulose is suitable for making cell walls of plants.

2.1 Introduction to cells

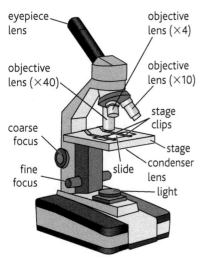

eyepiece lens

objective lens (×4)

objective lens (×40)

objective lens (×10)

stage clips

coarse focus

stage

fine focus

slide

condenser lens

light

Figure 2.1.1 *A typical school or college light microscope (the condenser lens is beneath the stage)*

☑ Study focus

$1\,nm = 1 \times 10^{-6}\,mm$; there are $10^3\,nm$ in a µm; there are $10^3\,µm$ in a mm.

☑ Study focus

When calculating magnifications and actual sizes always take measurements from photographs or drawings/diagrams in millimetres, never centimetres.

Using a light microscope

We use microscopes in biology because many of the things that we want to see are so small. There are two types of microscope that you need to know about:

- light microscopes
- electron microscopes.

The principles of the two types of microscope are similar. Light or electrons are focused by lenses on an object to be viewed. The light or electrons pass through or are absorbed or diffracted by the object. Further lenses focus the light or beam of electrons so that an image of the object can be viewed. In a light microscope such as that in Figure 2.1.1, there is a condenser lens, objective lenses and an eyepiece lens. The objective lenses give the microscope the ability to magnify the object by ×4, ×10 or ×40. The image formed by these lenses is magnified by the eyepiece lens, which is usually ×10. The overall magnifications are ×40 (low power), ×100 (medium power) and ×400 (high power).

No matter how much an object is magnified in the light microscope, it is impossible to see very fine detail. This is because the **resolution** of the microscope is limited by the wavelength of light. Resolution is the ability to see detail. The resolution of your eye is about 0.2 mm (200 µm), which means that any objects closer together than this are seen as one object not two. You cannot see any object smaller than 200 µm. The resolution of your eye is a function of the density of light receptor cells in your retina.

The resolving power of a microscope is dependent on the wavelength of light and how the light is interrupted by objects in the specimen. Visible light has a wavelength between 400 nm and 700 nm. Objects which are about half the size interrupt the light rays and can be seen in the light microscope. Anything that is smaller than 200 nm cannot be seen. The light microscope has a resolution of 200 nm (0.0002 mm), which means that two points separated by this distance can be seen as separate objects. Anything smaller, such as cell membranes, cannot be seen. If images in the light microscope are magnified then small objects larger than 200 nm in size are resolved and we can see them directly through the eyepiece and in photographs taken through the eyepiece.

Magnification is the ratio between the actual size of an object and the size of an image, such as a drawing or a photograph. Remember these rules for calculating magnifications and actual sizes:

$$\text{magnification} = \frac{\text{size of image}}{\text{actual size}}$$

$$\text{actual size} = \frac{\text{size of image}}{\text{magnification}}$$

Drawing cells

You can make temporary microscope preparations of plant and animal cells. Remove a leaf from a water plant, such as *Elodea*. Place on a slide in water and stain with iodine solution. Put a coverslip on top and dry the top and bottom surfaces of the slide. Place each slide in turn on the stage of the microscope and focus with the low power objective lens (e.g. ×4). When in focus, change to the medium power lens and then the high power lens.

Figure 2.1.2 shows drawings of these two types of cell made under high power.

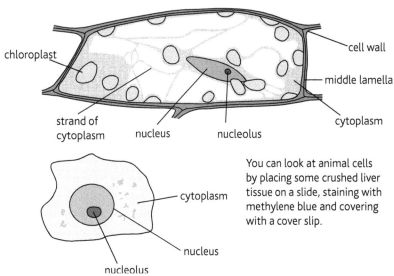

You can look at animal cells by placing some crushed liver tissue on a slide, staining with methylene blue and covering with a cover slip.

Figure 2.1.2 *Drawings of a leaf cell of* Elodea *and a liver cell (magnification = × 1000)*

When you make a high power drawing of cells, follow these simple rules:

- make the drawing fill at least half the space provided; leave space for labels and annotations (notes)
- use a sharp pencil (e.g. HB) and never use a pen
- use clear, continuous lines to show the outline of the cells and any internal details
- make sure that the proportions are the same as the cells you are drawing, e.g. you have the same width : length ratio
- show details of the contents of cells – draw what you see **not** what you know should be present; do not use any shading or colouring for the cell contents.

The similarities and differences between the plant and animal cells in Figure 2.1.2 are summarised in this table.

Feature	Plant cell	Animal cell	Feature	Plant cell	Animal cell
cell wall	✓	✗	chloroplast	✓	✗
fixed shape	✓	✗	large vacuole	✓	✗
cytoplasm	✓	✓	approximate size/μm	40	20
nucleus	✓	✓			

Summary questions

1   Describe how to make a temporary slide of plant tissue and focus it with the high power of a microscope.

2   Find out why microscopists use phase contrast and fluorescence microscopy.

3   Explain the difference between the terms *magnification* and *resolution*.

4   Calculate the actual sizes of the cells shown in Figure 2.1.2. Give your answers to the nearest micrometre.

∞ Link:

Look at page 35 for advice on plan drawings, you may be asked to make plan drawings and high power drawings of cells from the same microscope slide.

☑ Study focus

Cell membranes are too small to be resolved in the light microscope. They are often indicated by labels on drawings although they are not visible – if you put 'cell membrane' as a label what you are labelling is the outer boundary of the cell.

☑ Study focus

In Summary question 4, measure the length and width of the cells in millimetres and divide by the magnification. Remember to multiply by 1000 to give a result in micrometres and round up or down to the nearest whole number.

Cells in more detail

The light microscope does not show the details of cells. For this we need a microscope with a far greater resolution. A beam of electrons has a wavelength of 1.0 nm and therefore a resolving power of 0.5 nm which is 400 times better than that of a light microscope. This allows us to see a range of sub-cellular structures and even large molecules in photographs that are taken in the electron microscope.

Figure 2.2.1 shows an electron micrograph of a human white blood cell and Figure 2.2.2 a drawing made from the micrograph.

The electron micrographs of the plant and animal cells were made with a **transmission electron microscope**. The electron beam passes through the specimen to hit photographic paper to make the image. The paper can be replaced with a fluorescent screen or digial camera so that the operator can view the image and search for suitable features to photograph.

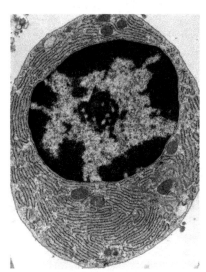

Figure 2.2.1 *Electron micrograph of a white blood cell (× 6700)*

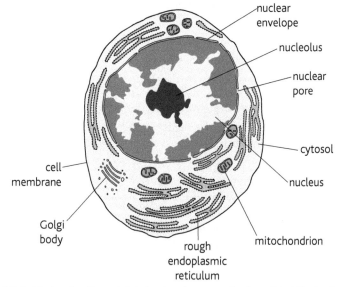

Figure 2.2.2 *Made using Figure 2.2.1 showing the organelles in a white blood cell (× 6700)*

The electron beam is focused by magnetic lenses in a column that contains the specimen. Pumps are used to remove the air from the column so that the electron beam is not deflected on its pathway through the specimen to the photographic paper, digital camera or fluorescent screen.

⊖⊖ Link:

An electron micrograph of a plant cell is on page 30.

The following table lists the differences between light microscopy and electron microscopy.

Feature	Light microscope	Electron microscope
wavelength/nm	400–700	1.0
resolution/nm	200	0.5
highest magnification without loss of detail	×1000 (at best: 1500)	×250 000 (or more)

✓ *Study focus*

Use the internet to find electron micrographs of cells. Use these to learn how to recognise the organelles listed on page 31.

There are advantages of using both light and electron microscopes to study cells.

- Light microscopes can be used to see living cells and living processes, such as movement of cells, phagocytosis, intracellular digestion, movement of chromosomes in nuclear division (mitosis and meiosis), division of cells and movement of cytoplasm.
- *The specimens observed in electron microscopes are dead as the processes of fixation and staining kills them; they are observed in a vacuum which means that all water must be removed as part of the preparation.*
- Various techniques, such as phase contrast, can be used to view living cells in light microscopes without the use of the fixatives and stains required in electron microscopy to deflect electron beams.
- *Some of the stains used in electron microscopy are heavy metals that combine with proteins and may distort cell structures to cause artefacts – objects that are not in the living cells.*
- Light microscopes can be used to see quite thick specimens.
- *The electron beam cannot penetrate thick specimens, so the specimens must be cut very thin (100–500 nm). It is easy to miss structures when cutting the sections this thin.*
- Images in the light microscope are in colour, either the natural colours of the specimen or the colours of stains (e.g. iodine, methylene blue and many specialised stains for prepared slides).
- *All images in the electron microscope are in black and white; computers are used to produce colour-enhanced images.*
- Light microscopes are cheaper to buy than electron microscopes, which also need trained technicians to operate and maintain them.
- Eukaryotic ribosomes are only 25–30 nm across, so they are below the resolution of light microscopes.
- *Electron microscopes allow viewing of very small structures within cells, such as ribosomes, that are not visible in the light microscope.*

✓ *Study focus*

Make sure you include a column for the features you choose in Summary question 4.

Summary questions

1 Calculate the actual length of the cell shown in Figure 2.2.1.

2 Explain the advantages of using the electron microscope to view cells, such as white blood cells and palisade mesophyll cells.

3 List the organelles visible in the electron micrographs of: **i** the white blood cell; **ii** the palisade mesophyll cell (see page 30 for the palisade mesophyll cell).

4 Make a table to show the **differences** between the animal and plant cells shown in the figures in this section and on page 30.

5 Suggest an investigation in which: **i** the electron microscope is best; **ii** the light microscope is best.

2.3 Cells and organelles

Learning outcomes

On completion of this section, you should be able to:

- define the term *organelle*
- list the organelles in plant and animal cells
- outline the functions of the organelles.

☑ Study focus

Think of a eukaryotic cell as a complex factory in which energy is generated, information stored, raw materials gained, products made and exported, substances and components moved about and waste removed. It is also a factory that can reproduce itself.

☑ Study focus

Make large diagrams of plant and animal cells with all the organelles that occur in these cells. Use the table opposite to label and annotate your diagrams. Keep adding more information about cells as you work through the rest of the book.

Cells have a variety of functions to carry out:

- obtain energy and convert it into usable form
- gain raw materials from their surroundings
- produce biological molecules, such as carbohydrates, lipids, proteins and nucleic acids
- package materials so they can be exported from the cell
- excrete waste materials
- store and retrieve genetic information.

You can see from the electron micrographs in the previous section that cells are divided into compartments. The sub-cellular structures labelled in the plant cell and in the animal cell are organelles. The functions of the cell are localised into these compartments because the conditions that the processes require are different. For example, the lysosomes in a white blood cell contain enzymes that digest all the biological molecules. These must be kept in a membrane-bound structure otherwise they would destroy the whole cell. Lysosomes are required to break down the bacteria that the white blood cell takes up into vacuoles.

The table shows all the organelles that you need to know about.
* = clearly visible in school/college light microscope.

Note that S stands for Svedberg unit which is used to measure small cellular structures and macromolecules. It refers to how they sediment when spun in a centrifuge.

Sub-cellular structures are made of

- membranes – examples are mitochondria and chloroplasts
- protein fibres – examples are centrioles.

The hair-like surface structures known as cilia (singular: cilium) and flagella (singular: flagellum) are also made of protein fibres.

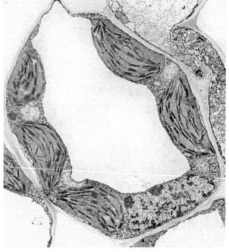

Figure 2.3.1 *Electron micrograph of a palisade mesophyll cell from a leaf. You can see five chloroplasts, a nucleus and a large central vacuole. Between the chloroplasts are two mitochondria that are smaller and paler. (× 3500)*

Organelle	Features	Function(s)
rough endoplasmic reticulum (RER)	flat sacs of membrane enclosing fluid filled space outer surface is covered in ribosomes	ribosomes carry out protein synthesis; RER transports proteins to Golgi body
smooth endoplasmic reticulum (SER)	like RER but with no ribosomes on outer surface	makes triglycerides (fats), phospholipids, cholesterol
Golgi body (or Golgi apparatus)	pile of flat sacs with vesicles forming around the edge	modifies and packages proteins; makes secretory vesicles and lysosomes
mitochondria (singular: mitochondrion)	formed of two membranes surrounding a fluid filled matrix inner membrane is highly folded to give large surface area for enzymes for respiration	site of aerobic respiration
ribosomes	attached to RER or free in cytoplasm – made of protein and RNA (30 nm in diameter; 80S)	assemble amino acids to make proteins
lysosomes	single membrane surrounds fluid filled with enzymes	contain enzymes for destroying worn out parts of cells and for digesting food particles and bacteria
*chloroplasts (plant cells only)	many internal membranes giving large surface area for chlorophyll, other pigments and enzymes	site of all the reactions of photosynthesis
cell membrane (or cell surface membrane)	bilayer of phospholipid with proteins (see page 36 for features)	cell boundary; retains cell contents; controls exchange of substances with surroundings
nuclear envelope	structure like that of ER with ribosomes on outer surface; pores to allow substances to pass between cytoplasm and nucleus	separates nucleus from cytoplasm; allows movement between the two
*nucleus	clearly visible in LM and EM when stained	contains store of genetic information as DNA in chromosomes
*nucleolus	darkly staining area in nucleus	produces ribosomes
centrioles (animal cells only)	made of protein fibres, structure is similar to the base of a cilium/flagellum	assemble the spindle to move chromosomes when nuclei divide

Summary questions

1 Make a poster that includes drawings of each of the organelles, their functions and roles within the cell. You may wish to make the poster in the form of a large table. You can add information to this poster as you work your way through the course.

2 Name the organelles that are involved in; **i** the synthesis and export of proteins from the cell; **ii** energy transformations; **iii** the intake and digestion of materials by white blood cells.

3 Describe how the following pairs of organelles differ from each other in structure and function:
i mitochondria and chloroplasts; **ii** nucleus and Golgi body; **iii** smooth and rough ER; **iv** cell membrane and nuclear envelope.

4 Make a list of the functions that occur in cells. Name the organelles that are involved in each of the functions that you have listed.

2.4 Eukaryotes and prokaryotes

Learning outcomes

On completion of this section, you should be able to:

■ identify eukaryotic and prokaryotic cells in the light and electron microscopes

■ describe the structure of a prokaryotic cell

■ state the differences between prokaryotic and eukaryotic cells

■ outline the endosymbiotic development of eukaryotic cells.

∞ Link

Look at the diagram and the table to see how prokaryotic cells carry out the functions you identified in Section 2.3, Summary question 4, on pages 30 to 31.

The cells described in the previous pages are **eukaryotic** cells. This is because they have a nucleus, which is the meaning of the term **eukaryote**. These cells have complex membrane systems that subdivide the cell into organelles. Organisms such as bacteria do not have this sort of cell structure. This much simpler cell structure is called **prokaryotic**. Figure 2.4.1 shows a diagram of the features seen in **prokaryotes**. No one cell has all these features – for example, plenty of bacteria do not have capsules or flagella.

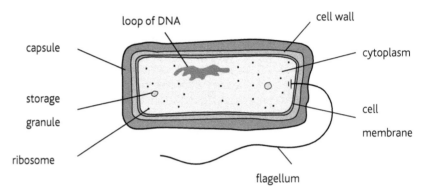

Figure 2.4.1 The features of prokaryotic cells

Structure	Features	Functions
capsule	thick, compact, slimy layer outside cell wall	provides protection for parasitic bacteria, e.g. against phagocytes reduces chances of dehydrating
cell wall	made from murein, not cellulose; murein is a polysaccharide cross linked into mesh by peptides	prevents cells from bursting when in solutions of higher water potential
cell (surface) membrane	phospholipid bilayer with proteins; no cholesterol	partially permeable to allow entry and exit of substances
cytoplasm	contains ribosomes, storage granules, enzymes; contains bacterial chromosome	respiration, protein synthesis, other chemical reactions
ribosome	smaller than eukaryotic: 20 nm in diameter; 70 S (note prokaryotes have no ER)	protein synthesis
storage granules	stores of glycogen, lipids and phosphate	store of energy and lipids for membrane synthesis
flagella (singular flagellum)	made of one fibre and not enclosed by membrane	base rotates to give corkscrew motion; moves bacterium through liquids
genetic material	bacterial chromosome is loop of DNA; plasmids are smaller DNA loops; both are in the cytoplasm (prokaryotes have no nuclear envelope)	stores of genetic information; one copy of each gene, so prokaryotes are haploid (see page 80)

The differences between prokaryotic and eukaryotic cells are listed in this table.

Feature	Prokaryotic cell	Eukaryotic cells	
		Plant cells	Animal cells
typical size/μm	0.5–3.0	40–60	20
capsule	found in some	✗	✗
cell wall	✓ (murein, not cellulose)	✓ (made of cellulose)	✗
membrane-bound organelles	✗	✓	✓
ribosomes	smaller: 20 nm/70 S	30 nm/80 S	30 nm/80 S
nucleus	✗	✓	✓
DNA	bacterial chromosome is a loop of DNA in cytoplasm	chromosomes made of linear DNA found within nuclear envelope	

Endosymbiosis

Endosymbiosis is the idea that organelles have evolved from prokaryotes. Mitochondria and chloroplasts share many features with prokaryotes. They both have:

- a loop of DNA which is similar to the bacterial chromosome, although some genes that control features in chloroplasts and mitochondria are in chromosomes in the nucleus
- small 70 S ribosomes
- transfer RNA molecules that only function within the organelles and are coded for by genes in the mitochondrial and chloroplast DNA.

The similarities suggest that these organelles have been derived from bacteria as shown in Figure 2.4.2.

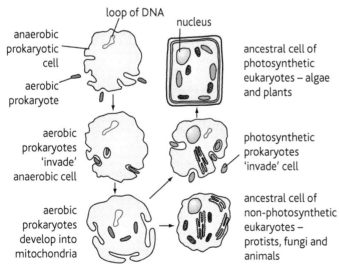

Figure 2.4.2 Endosymbiosis

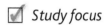 **Study focus**

Symbiosis means that two species interact with each other and both gain benefit from the association. In endosymbiosis, cells live inside other cells in a mutual beneficial arrangement. This led, so it is thought, to prokaryotes evolving into mitochondria and chloroplasts.

Summary questions

1 Define the terms *eukaryote* and *prokaryote*.

2 List the structures found in:
   i both prokaryotic cells and eukaryotic cells; **ii** only in prokaryotic cells; **iii** only in eukaryotic cells.

3 Define the term *endosymbiosis* and outline the evidence that supports the idea.

4 Find out about Lyn Margulis, one of the early champions of the idea that organelles, such as mitochondria and chloroplasts, evolved by endosymbiosis.

On completion of this section, you should be able to:

- define the terms *tissue* and *organ*
- list the differences between cells, tissues, organs and organ systems
- list the major organs in plants and animals
- make a plan drawing of a cross section of a plant organ.

☑ Study focus

The outer layer of the skin is an epithelium. Epithelia line the inner surfaces of organs, such as the intestine, the oviduct and the trachea.

Some eukaryotic organisms are unicellular consisting of single cells. There are many different species of these and they are classified as protists. All the body processes of a protist occur within one mass of cytoplasm with a nucleus. Many eukaryotes have bodies divided into cells, they are described as multicellular. Cells specialise in different functions and are organised together to give sheets or masses of cells. Groups of similar cells are known as tissues. The table includes some examples of epithelial cells that forms sheets of cells.

Type of epithelial cell	Shape	Single or multi-layered	Surface features	Example of location
squamous		single	none	alveoli in the lungs
		multi-layered	none	epidermis in the skin
cuboidal		single	microvilli – short, thin extensions of cell membrane to increase surface area	lining kidney tubules
columnar		single	microvilli	lining small intestine
		single	cilia	lining trachea and bronchus, lining oviduct

Tissues are groups of similar cells that carry out the same function.

Organs are structures composed of different tissues that carry out one or several major functions for the body.

The table shows you examples of plant and animal tissues and organs.

Plant tissues	Plant organs	Animal tissues	Animal organs
epidermis chlorenchyma sclerenchyma parenchyma xylem phloem	leaf stem root	epithelium muscle blood bone cartilage adipose	brain, spinal cord, tongue, eye and ear individual muscles, e.g. biceps individual bones, e.g. femur liver, pancreas, stomach, small and large intestine, kidney, ovary, testes, uterus

In animals, **organ systems** comprise of several organs that work together to bring about one major function of the body. The excretory, nervous, sensory, muscular, reproductive, digestive and endocrine systems are examples of organ systems in mammals.

You will have to make **plan drawings** of sections through organs to show the distribution of tissues. Figure 2.5.1 shows a transverse section through the root of *Ranunculus*. Figure 2.5.2 shows a plan drawing made from the section.

☑ *Study focus*

Practise making plan drawings of images of cross sections of roots, stems and leaves that you can download.

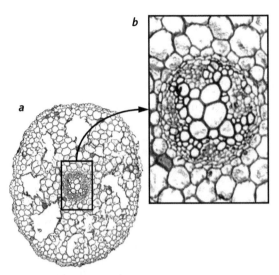

Figure 2.5.1 a *Transverse section of a root of* Ranunculus *with* **b** *an enlarged view of the transport tissue (xylem and phloem)*

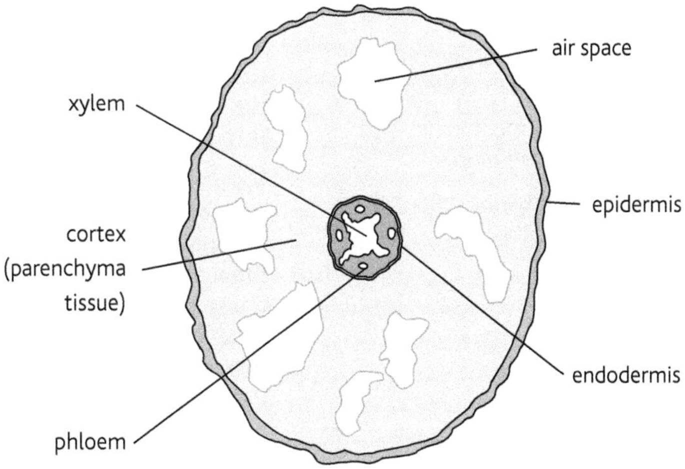

Figure 2.5.2 *A plan drawing made from the section of the root of* Ranunculus

Follow this guidance when making plan drawings:

- make the drawing fill at least half the space provided; leave space around the drawing for labels and annotations
- use a sharp pencil (e.g. HB) and never use a pen
- use thin, single, continuous lines to show the outline of the organ and the boundaries of the tissues
- make the proportions of the tissues in the plan drawing exactly the same as in the section
- do **not** include drawings of cells and do **not** shade or use any colours.

Link

TS is the abbreviation for transverse section (also called a cross section) and LS is the abbreviation for longitudinal section.

Summary questions

1 Define the following terms: *tissue*, *organ* and *organ system*.

2 List the organs that comprise the following organ systems in humans: male reproductive system, female reproductive system, digestive and nervous.

3 Find out and list the functions of the following animal tissues: muscle, blood, skeletal, epithelial.

4 Find out and list the functions of the following plant organs and tissues: leaf, stem, root, xylem, phloem, parenchyma, chlorenchyma, sclerenchyma and epidermis.

5 State the purpose of making plan drawings of sections of organs.

6 Suggest some advantages of being multicellular rather than unicellular.

Learning outcomes

On completion of this section, you should be able to:

- list the components of cell membranes
- describe the arrangement of components in a cell membrane
- explain why membranes are known as fluid mosaic
- state and describe the functions of the components of cell membranes.

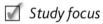 Study focus

Figure 2.6.1 shows a simple diagram that you can copy and learn. You may have to draw it in Paper 2.

⚬⚬ Link

Follow the advice given on page *v* about finding other diagrams and animations of cell surface membranes on the internet.

Eukaryotic cells are composed of membranes that provide a barrier between the contents of a cell and its surroundings. They also form compartments inside the cell that are the organelles described on page 31.

The main components of membranes are phospholipids and proteins. The arrangement of these molecules is known as a **fluid mosaic** with proteins floating in a sea of lipid. Phospholipids form a bilayer with a hydrophobic interior and hydrophilic surfaces facing the cytoplasm and the external surroundings. Some protein molecules are associated with the inside or outside of the membrane, but transmembrane proteins pass through the membrane forming pores and channels.

Figure 2.6.1 gives the false impression that the components of a membrane are fixed in position and do not move about. In fact, the phospholipids are in constant motion moving about within each monolayer, which is why the membrane is described as fluid. Some phospholipid molecules 'flip' from one monolayer to the other, but generally they remain within their monolayer and the composition of the two monolayers is different in order to face the different environments. Proteins move about within the lipid. The proteins of epithelial cells are rarely able to move over the whole surface of the cell, but remain on one surface. For example, carrier proteins for glucose and sodium are restricted to the microvilli on the surface of epithelia in the small intestine. They are not found on the side of the epithelial cells that faces capillaries.

Component of cell membrane	Function
phospholipids	- form a bilayer that acts as a barrier between exterior of cells and cytoplasm - fluid, so proteins can move about - permeable to non-polar molecules - impermeable to ions and large polar molecules
proteins	- transmembrane proteins are channels and carriers
glycoproteins – proteins with short chains of sugars attached on exterior side	- receptors for hormones, neurotransmitters at synapses between nerve cells and between nerve cells and muscle cells - recognition sites for other similar cells when forming tissues during development - recognition sites for antibodies and lymphocytes
glycolipids – lipids with one chain of sugars attached on exterior side	
cholesterol	- stabilises the phospholipid bilayer by binding to polar 'heads' and non-polar 'tails' - controls fluidity by preventing phospholipids solidifying at low temperatures and becoming too fluid at high temperatures

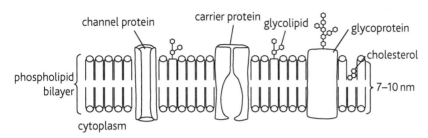

Functions of membranes

At cell surfaces:

Function	Comments
barrier	▪ many water-soluble substances cannot pass across; large molecules that are required in the cell, such as proteins, cannot leave
permeability	▪ partially permeable (**not** semi-permeable) as some substances can pass through phospholipid bilayer and others through transmembrane proteins; many substances cannot pass through
absorption through membrane	▪ cell membranes extended into microvilli to increase surface area for absorption through phospholipid bilayer and transmembrane proteins
movement by bulk flow	▪ substances that cannot pass through the bilayer or transmembrane proteins are carried to or from the cell membrane in small vesicles or larger vacuoles. See endocytosis and exocytosis on page 39
recognition	▪ receptors have binding sites for cell-signalling molecules, such as hormones and growth factors

Within cells:

Function	Comments
form compartments	▪ isolation of hydrolytic enzymes in lysosomes so they do not harm the cell ▪ concentration of substances, e.g. enzymes and their substrates ▪ provision of areas with different pH, e.g. interiors of chloroplast and mitochondria have different pH from rest of cytoplasm
provide large surface area	▪ chloroplasts and mitochondria have membranes for forming ATP ▪ chloroplasts have membranes for many pigment molecules, such as chlorophyll
intracellular transport	▪ vacuoles and vesicles move substances from cell surface membrane into the cell, from RER to Golgi body and Golgi body to cell surface membrane

Make a diagram of an LS through a columnar epithelial cell from the small intestine. Draw microvilli on one surface. Draw in the organelles you would expect to find in this cell. Use preceding Sections 2.2, 2.3 and 2.5 to help you.

Summary questions

1 Define the following terms: *cell surface membrane; fluid mosaic; phospholipid bilayer; cholesterol; glycoprotein; glycolipid; transmembrane protein*.

2 Describe the distribution of membranes in a eukaryotic cell.

3 State the width of a cell membrane.

4 Explain why membranes are not visible in the light microscope.

5 Suggest why membranes are often visible as two parallel lines in electron micrographs of cells magnified by 100 000.

6 Find and summarise the evidence for the fluid mosaic structure of cell membranes.

7 Explain why plant cells do not have microvilli.

Did you know?

Aquaporins are special channel proteins with a polar lining to allow water to diffuse in or out of cells.

☑ Study focus

This diagram shows the different concentrations of substances on either side of the membrane.

∞ Link

Osmosis and water potential are focused upon in more detail on page 40.

Cell membranes are barriers to movement, but they also allow substances to move in and out of cells. Substances enter because cells need them – they are their raw materials. Waste substances move out because they may be toxic and cell products leave to be used elsewhere or because they function outside the cell.

Some substances are small enough to pass through the membrane, some are too big and must be enclosed within a membrane in vacuoles or vesicles.

Movement is either passive or active.

Passive movement

Molecules cross membranes by diffusion down their concentration gradient. The cell does not need to use its own energy to move the molecules. The energy comes from the kinetic energy of the molecules.

Metabolic energy may be used to create the gradient in the first place – for example, movement of oxygen from the air in alveoli in the lungs into the blood relies on a gradient maintained by breathing and the flow of blood from the heart.

■ Simple diffusion. Molecules pass through the phospholipid bilayer. Fat soluble substances such as ethanol pass readily through the bilayer. Oxygen and carbon dioxide are small enough to pass through as both are small and uncharged.

■ Facilitated diffusion. Polar molecules cannot pass through the phospholipid bilayer so there are channel proteins for them. These are open all the time to allow movement down a gradient. Carrier proteins also allow movement, but these open when molecules bind to them, they are not open all the time. Water molecules are polar and they cannot easily pass through the bilayer (see page 36).

■ Osmosis. This is a special type of diffusion that involves the movement of water through a partially permeable membrane from a place with a higher water potential to a place with a lower water potential.

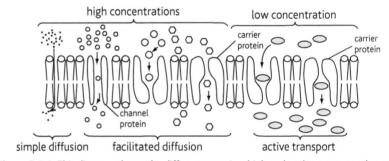

Figure 2.7.1 *This diagram shows the different ways in which molecules cross membranes*

Water potential

Water potential is the tendency of water to move from one place to another and is determined by:

■ the quantity of water present

■ the concentration of solutes, such as ions and sugars

the pressure exerted by the cell wall (in plant cells and prokaryotes, not in animal cells).

Solutions with high water potentials have low concentrations of solute molecules. Solutions with low water potentials have high concentrations of solute molecules.

Water moves from a solution with a high water potential to a solution with a low water potential. Cells contain cytoplasm and plant cells also contain cell sap inside their large vacuoles. These contain many solutes, such as proteins, sugars and ions, and thus cells have a much lower water potential compared with distilled water or tap water. Water molecules always travel back and forth across membranes. When cells are placed into distilled water, more water molecules move into the cell down the water potential gradient than move out. The *net* movement of water is into the cell. When placed into sea water, the movement in and out of the cell is usually about the same: there is no *net* movement of water molecules. The water potential of cell contents is about the same as the water potential of sea water.

Active movement

- Active transport. Some substances are required inside cells that are present in very low concentrations outside. They will not move into cells by diffusion. Instead they will tend to move out! To gain these substances they must be moved against their concentration gradients. Cells have to use their own energy released by respiration to move these substances. Carrier proteins accept molecules on one side of the membrane, change shape to move the molecules from one side to the other, then release them. This happens in root cells that absorb ions from soil water and in the kidneys to absorb glucose so it is not lost in urine.

- Bulk transport. Cells take in particles from their surroundings by endocytosis. For example, phagocytic while blood cells take in bacteria. These are much too large to pass through the transmembrane proteins or through the bilayer so are enclosed in a vacuole made of cell surface membrane. This reduces the quantity of membrane at the cell surface. Lymphocytes produce antibodies that are large molecules. These are enclosed within vesicles made by the Golgi body and move to the cell surface membrane. They fuse with the membrane so that their contents are expelled to the outside by exocytosis. The vesicle membrane becomes part of the cell surface membrane. Movement of vesicles and vacuoles requires energy from the cell.

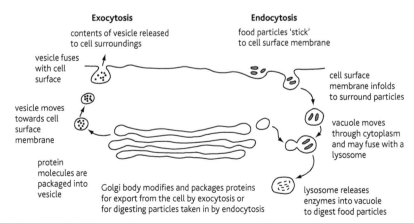

Figure 2.7.2 Bulk transport across membranes – exocytosis and endocytosis

 Study focus

Channel and carrier proteins are transmembrane proteins. Channel proteins are used in facilitated diffusion; carrier proteins in facilitated diffusion and active transport.

Study focus

Vesicles or vacuoles? Vesicle is a term used to describe small vacuoles.

Summary questions

1 Define the following terms: *diffusion, simple diffusion, facilitated diffusion, vesicle, vacuole, active transport, exocytosis, endocytosis, secretion.*

2 Explain why substances cross membranes in and out of cells.

3 Explain why there are special channel proteins (aquaporins) for water.

4 Explain the differences between the following pairs: **i** channel protein and carrier protein; **ii** simple and facilitated diffusion; **iii** passive and active transport across membranes; **iv** secretion and excretion; **v** exocytosis and endocytosis.

5 Find examples of endocytosis and exocytosis and list them.

Learning outcomes

On completion of this section, you should be able to:

- define the terms *water potential gradient*, *plasmolysis*, *turgid*, *flaccid*
- explain that osmosis is a special type of diffusion
- describe and explain the effects of immersing plant and animal tissues in solutions of different water potential.

Figure 2.8.1 *Sections cut from egg plants make good material to investigate osmosis and plant cells. The changes happen very quickly.*

✓ Study focus

The results in Investigations 1 and 2 are descriptive or qualitative. Try Summary question 2 on page 43 using your knowledge of osmosis and plant and animal cells.

✓ Study focus

Do not refer to 'weak' and 'strong' solutions. It is best to refer to their actual concentrations if they are given in a question or to say that a solution has a low or high concentration of the solute concerned.

Salty solutions

During your course you should develop your skills of experimental design, data presentation and data analysis and interpretation. This section deals with these skills in the context of investigating the effect of immersing tissues in solutions of different water potential.

Read through the four investigations described in this section and answer the questions that follow. There are more questions on this topic in Section 2.10.

Investigation 1

Cut the fruit of an egg plant, *Solanum melongena*, into equal-sized sections. Place the sections into five different concentrations of salt (sodium chloride). The results are summarised here:

Concentration of sodium chloride/mol dm^{-3}	Observations on the egg plant tissue compared with freshly cut tissue
0.00 (distilled water)	firmer
0.25	very similar
0.50	slightly softer
0.75	softer than in 0.50 mol dm^{-3}
1.00	very soft

Investigation 2

Add one drop of fresh blood to different concentrations of salt. Stir and observe the appearance of the contents of the test-tubes. Remove a sample and place on a microscope slide and observe the appearance of the red blood cells.

Concentration of sodium chloride/mol dm^{-3}	Observations of the blood
0.00 (distilled water)	no cells present, red solution
0.06	few cells present, red solution
0.15	cells present, no red solution
0.25	shrunken cells present, no red solution
0.50	shrunken cells present, no red solution

Investigation 3

This is an investigation to obtain some quantitative results using plant tissue.

Choose a suitable plant tissue such as the storage tissue from tubers of: European potato, *Solanum tuberosum*; sweet potato, *Ipomoea babatas*; or yam, *Dioscorea alata*. The part of these plants that you eat is storage tissue.

- Use a cork borer or a chip machine to cut the storage tissue into cylinders or 'chips'.
- Trim the ends so they are squared off and make the pieces all the same length.
- Weigh the pieces and record their individual masses.
- Place the pieces into separate test-tubes each containing solutions of sucrose as shown in the results table below.
- After 12 hours reweigh the pieces and record the results.

In this investigation it is necessary to calculate the percentage change in mass because the initial masses of the potato pieces are different. The percentage change in mass is the derived variable. It is calculated as follows:

$$\text{percentage change} = \frac{\text{change in mass}}{\text{original mass}} \times 100$$

Study focus

These are the variables in this investigation:

- independent – concentration of sucrose solution
- dependent – change in mass
- derived – percentage change in mass
- control – type of tissue, length of immersion, temperature, volume of solution.

Can you think of any more control variables? There is more about control variables on page 59.

Concentration of sucrose solution/mol dm⁻³	Initial mass/g	Final mass/g	Change in mass/g	Percentage change in mass	Mean percentage change in mass
0.0	1.26	1.49	0.23	18.3	19.4
	1.23	1.51	0.28	22.8	
	1.22	1.43	0.21	17.2	
0.2	1.22	1.31	0.09	7.4	5.1
	1.28	1.32	0.04	3.1	
	1.25	1.31	0.06	4.8	
0.4	1.43	1.29	−0.14	−9.8	−7.0
	1.32	1.26	−0.06	−4.5	
	1.34	1.25	−0.09	−6.7	
0.6	1.32	1.10	−0.22	−16.7	−15.2
	1.26	1.07	−0.19	−15.1	
	1.22	1.05	−0.17	−13.9	
0.8	1.28	0.98	−0.30	−23.4	−21.3
	1.25	1.01	−0.24	−19.2	
	1.23	0.97	−0.26	−21.1	
1.0	1.3	0.95	−0.35	−26.9	−23.1
	1.27	1.01	−0.26	−20.5	
	1.23	0.96	−0.27	−22.0	

Three pieces were used for each concentration to see how much variation there is in the results. If there is little variation in the results for each concentration then you can say that the results are reliable. The three pieces of potato cannot be labelled so they are put into separate test-tubes rather than all in the same container.

These results are plotted on the following graph.

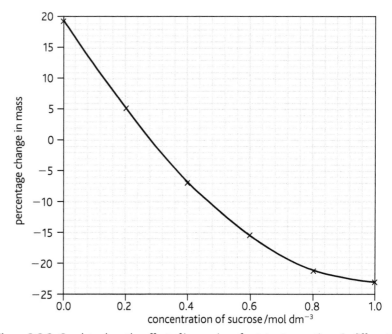

Figure 2.8.2 *Graph to show the effect of immersion of potato storage tissue in different solutions of sucrose*

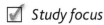

Link

Use the data on page 45 to plot a conversion graph so you can find the water potential in kPa of the potato storage tissue.

Study focus

When there is no overall movement of water, it is best to state that there is no net diffusion of water by osmosis.

The curve drawn on the graph passes through zero per cent at which there is no change in mass. This did not happen in any of the results, but we can use this intercept on the graph to find the concentration at which we could predict that it would happen. Use the intercept at zero per cent to find the concentration, which is 0.28 mol dm⁻³.

The changes in texture, length and mass are the result of movement of water between the tissues and the sucrose solution.

- In solutions more concentrated than 0.28 mol dm⁻³ water moves by osmosis out of the cells into the solution so the pieces of tissue decrease in size.

- In solutions less concentrated than 0.28 mol dm⁻³ water moves by osmosis from the surrounding solution into the cells so the pieces of tissue increase in size.

- In the solution in which there is no change in mass there is no overall movement of water. In this solution, water molecules are still able to move in and out of the cells through the cell surface membranes. The membranes do not suddenly become impermeable when put in a solution like this. So there is no net diffusion of water by osmosis.

Plasmolysis

Investigations 1 and 3 were carried out with plant tissue and it is difficult to see the changes that occur to the individual cells in the storage tissue as it was in the blood cells in Investigation 2. We can use epidermal tissue from onion scale leaves to see the changes, such as those in Investigation 4.

Investigation 4

Peel off the epidermis of an onion scale leaf and cut into pieces small enough to fit below a cover slip. Put the pieces into different solutions of sucrose or salt and leave them for 10 minutes. Put each piece on a microscope slide in the solution in which they were immersed. Observe under the microscope and look for cells like those in the photographs.

In Figure 2.8.3a, distilled water moves into each cell by osmosis to increase the volume of the vacuole, which pushes the cytoplasm and cell surface membrane against the cell wall. The cellulose cell wall withstands this turgor pressure as it has an equal and opposite force known as pressure potential. Cells filled with water like this are **turgid**.

In solutions with high concentrations of sucrose or salt, water moves out of the cells by osmosis. Most of the water comes from the vacuole that decreases in size, pulling the cytoplasm and cell membrane away from the cell wall. The space between the cell wall and the cell surface membrane fills with the external solution. This condition is known as **plasmolysis** and cells in this condition are plasmolysed. Cells like this have lost their turgidity and are **flaccid** (soft).

In distilled water all the cells are turgid. As the concentration of the salt solution increases the percentage of plasmolysed cells increases until at 0.5 mol dm^{-3} all cells are plasmolysed (see Figure 2.8.3b). At a certain concentration the cells are not plasmolysed, but the cell contents are not pushing against the cell wall and therefore the cell wall is not exerting any pressure potential. The water potential of this salt solution is equivalent to the water potential of the cell contents (cytoplasm and vacuole).

In Investigation 3, the storage tissue that was cut from the potato and put into the solutions was very firm. This means that the cells are turgid. However, over time the pieces placed in two of the solutions gained mass as water diffused into the cells (see 0.0 and 0.2 mol dm^{-3} in Figure 2.8.2). The cells were able to absorb some water to become more turgid. This means that the cells from the freshly cut tissue were not fully turgid as they have absorbed water. The pieces were left long enough for water to diffuse in or out and the tissues to reach equilibrium in the solutions.

If you take storage tissue that has been in water for 12 hours, weigh it and then put it back into water it will not show any further increase in mass as the cells are fully turgid and there is no space for any more water. The water potential of these cells is 0 kPa which means that their ability to absorb water is zero. The water potential of freshly cut tissue will be a negative figure as we have seen that the tissue absorbs water.

Some plants are adapted to live in dry soils and in salty soils; they absorb water from soils with a very low water potential. Water is absorbed passively by osmosis. To maintain a water potential gradient the root tissues must have water potentials lower than those of the soils.

When writing about osmosis and the movement of water between cells, or between cells and their environment, always refer to water potential. You may find the terms hypotonic, hypertonic and isotonic used in older text books and on websites. These terms describe the behaviour of cells in different solutions; they do not explain what happens to them. A hypotonic solution is one in which cells increase in volume; a hypertonic solution is one in which they decrease in volume and an isotonic solution is one in which they do not increase or decrease in volume. The changes in volume, or lack of them, can be explained in terms of the water potential gradient. If cells gain water by osmosis the water potential of the external solution is higher than that of the cell or tissue. If a cell decreases in volume the water potential of the cells is higher than that of the external solution. You can say that there is a water potential gradient into or out of the cells. In an isotonic solution there is no water potential gradient and that is why the cells remain the same volume.

☑ Study focus

When writing about osmosis and the movement of water between cells and in organisms always refer to water potential. It is best not to use the terms hypotonic, hypertonic or isotonic since they describe the behaviour of cells in different solutions, they do not *explain* what happens to them.

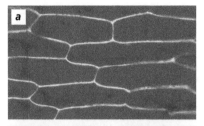

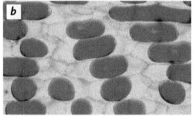

Figure 2.8.3 a *The cells are fully turgid as they are immersed in distilled water; **b** the cells are plasmolysed as they are immersed in a solution with a high concentration of salt*

Summary questions

1 Define the terms *osmosis*; *water potential gradient*; *turgid*, *turgor pressure*; *pressure potential*; *flaccid*; *plasmolysis*.

2 Explain, using the term *water potential*, the results obtained in investigations 1 and 2 on page 40.

3 Explain why animal cells: **i** do not show plasmolysis when in a concentrated salt solution; **ii** burst when placed into distilled water.

4 Plants, such as sea lavender and mangrove, live in very salty soil, which has a very low water potential. Explain how you would find the water potential of the root tissue of these plants.

Learning outcomes

On completion of this section, you should be able to:

- organise information about cells to show how the different parts of a cell help it to function efficiently

- use a graphic organiser to display information about cells and to link the structure of organelles to their functions

- use data provided to plot line graphs to find the water potentials of solutions of sucrose and sodium chloride.

Cells can be a rewarding topic to revise as it links to so many other topics. To begin with here are some ideas for activities to assist revising 'cells'.

Draw a large diagram of a plant cell and a large diagram of an animal cell. You may be able to use some A3 or poster-sized paper for this. Label the diagrams and annotate with the functions of the different structures that you have drawn. You can also include a drawing of a prokaryotic cell. You could try drawing them all to scale.

In some of the summary questions you will see we have asked you to differentiate between pairs of terms. This is a very common type of exam question, so practise doing this with the questions we have given you, but also with others.

Make your own glossary of the technical terms in this chapter and in Chapter 1. Many of the terms are in the glossary starting on page 178, but it is a good idea if you write your own with longer entries. Also write out sentences using the terms. You can test yourself by printing out terms and definitions separately and then matching them together. Or you can ask a friend or family member to read out the definitions and you call out the terms, or they can read out the terms and you provide the definitions. Some terms are very similar. Make sure that you spend time learning how to give precise definitions for them.

Make a graphic organiser of some sort. For example you could make a spider diagram to collect together all the facts and concepts you have covered in this chapter. See below for one to get you started. Write the word 'cells' in the middle of a piece of paper and then think of different aspects of the topic and write them down around the central word. Then make links to other topics so you finish with all the topics you have covered on one large sheet of paper. You should make links between different topics; these links will prepare you for answering questions that deal with more than one topic.

Print out some electron micrographs of cells so you learn to identify organelles from real cells, as well as from diagrams and drawings made from electron micrographs. Search online for images of organelles and different types of specialised cell.

Think of cell membranes as dynamic structures, not static. Search online for animations of cell membranes that show the movement of phospholipids and proteins and also the movement of substances across membranes by simple diffusion, facilitated diffusion, active transport, osmosis and bulk flow.

There are many different types of graphic organiser. Try searching for examples of mind maps, concept maps, spider diagrams and flow charts, and also find instructions about how to make them. Use a variety of different graphic organisers to help your revision.

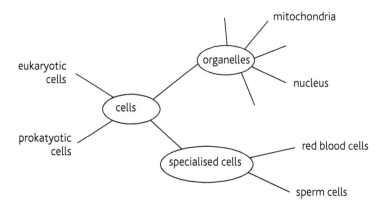

Make a poster of cell structure incorporating all the most important information. You can use it to help your revision.

Find diagrams or photographs of specialised cells and identify how their structure is related to their function. Here are some examples that you could research:

- red blood cell
- pancreatic exocrine cell
- pollen grain
- human sperm cell
- human ovum
- palisade cell
- guard cell
- white blood cell – phagocyte
- white blood cell – lymphocyte

∞ *Link*

You need the graph that converts sucrose concentrations to water potentials to find the water potential of the potato storage tissue from Investigation 3 on page 40.

Use the information you find to make a table with the following headings:

Name of cell; Tissue; Organ; Organ system; Structural features; Function.

To do this you will probably have to make the table in landscape format. Use the information in your table to explain how the structure of each cell is related to its function.

Plot conversion graphs so that you can find the water potential of sodium chloride and sucrose solutions. You will need the conversion graph for sodium chloride when answering Questions 8 and 9 on page 47. Note that 1 MPa = 1000 kPa

Concentration of sodium chloride/mol dm^{-3}	Water potential/ MPa	Concentration of sucrose/mol dm^{-3}	Water potential/ MPa
0	0	0	0
0.1	−0.440	0.10	−0.260
0.2	−0.850	0.20	−0.540
0.3	−1.260	0.30	−0.860
0.4	−1.680	0.40	−1.120
0.5	−2.120	0.50	−1.450
0.6	−2.570	0.60	−1.800
0.7	−2.970	0.70	−2.180
0.8	−3.420	0.80	−2.580
0.9	−3.860	0.90	−3.000
1.0	−4.320	1.00	−3.500

On page 22 we introduced you to multiple-choice questions which have one correct answer. But there is another type of multiple-choice question that you can expect on your CAPE Biology Paper 1. This type has one, two, three or even four answers and you have to decide which combination is correct. Here are some examples.

1 The following are found on the surfaces of cells.
   1 microvilli
   2 cilia
   3 receptor proteins
   4 cellulose cell wall
   5 glycoproteins

   Which of the following are NOT found on the surfaces of animal cells?
   A 1 and 2
   B 3 and 5
   C 1 and 3
   D 4 only.   (K)

Read the list carefully. The first two are structures that protrude from the surface – microvilli to increase the surface area; cilia to bring about movement. Receptor proteins and glycoproteins are part of the cell surface membrane and interact with molecules from the outside environment of the cell. Cellulose cell walls are only found surrounding the cell surface membrane of plant cells so D is the right answer.

2 Four plant cells from the cortex of a root have the following water potentials.

cell	1	2	3	4
water potential/MPa	−0.76	−1.34	−0.65	−1.01

The cells are all in contact as shown in Figure 2.10.1.

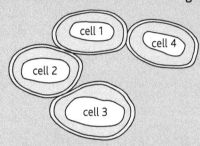

Figure 2.10.1

In which direction will there be movement of water between the cells?
A 3 to 4 and 2
B 2 to 1 and 3
C 1 to 4 and 2
D 4 to 2 and 1   (U of K)

To answer this question you need to know **three** things:
- as you add more solute to a solution the water potential decreases
- water moves down a water potential gradient from a region of higher water potential to a region of lower water potential
- water potentials are given negative numbers, so −0.65 is the highest water potential and −1.34 is the lowest.

Water can only move between the cells that are touching each other, as is the case with two choices you are given in the question. Cell 1 has high water potential and the water potential of the neighbouring cells are lower than −0.76 MPa, so C must be the right answer. Check B and you will find water is going *against* the water potential gradient, which cannot happen.

Water also moves from Cell 3 to Cell 2, but that is not one of the options. When answering this type of question you should anotate the diagram to help you work out the answer. Here you could draw arrows between the cells.

Try these questions for yourself as exam practice. These are all SAQs.

3   This is a list of organelles (A to H) and a list of cell functions (1 to 8). Match the organelles with their functions.

A	cell surface membrane	1	storage of genetic information
B	Golgi body	2	endocytosis
C	rough endoplasmic reticulum	3	formation of secretory vesicles
D	nucleus	4	formation of proteins
E	lysosome	5	formation of lipids
F	smooth endoplasmic reticulum	6	respiration
G	nucleolus	7	production of ribosomes
H	mitochondrion	8	storage of hydrolytic enzymes

4   State THREE structural features of prokaryotic cells that are not found in eukaryotic cells.

5   a   Draw a diagram of a mitochondrion that is at least 50 mm in length.
    b   Assume the actual length of the mitochondrion is 3.5 μm. Calculate the magnification of your drawing.
    c   Explain why:
        i   some cells have large numbers of mitochondria
        ii   fertilised eggs without any functional mitochondria do not survive.
    d   Explain why mitochondria are thought to have developed from prokaryotic cells.

6   a   Draw a diagram of a Golgi body that is at least 50 mm across.
    b   Show on your diagram how a Golgi body forms secretory vesicles.
    c   Describe what happens to secretory vesicles containing substances that are exported from cells.
    d   State one function of the Golgi body other than production of secretory vesicles.

7   a   Make a simple drawing to show the structure of a cell surface membrane.
    b   Label the components of the membrane.
    c   State the width of the membrane.
    d   Use your diagram to explain how polar molecules pass across the membrane.

8   The table shows the results of counting the number of red blood cells after a sample of blood is placed into different solutions.

Concentration of sodium chloride solution/mol dm^{-3}	Percentage of red blood cells destroyed by haemolysis
0.00	100
0.02	100
0.04	100
0.06	80
0.08	45
0.10	28
0.12	12
0.14	0
0.16	0
0.18	0
0.20	0
0.22	0
0.24	0

   a   Plot a graph of these results.
   b   State the water potential at which 50% of the cells are destroyed by haemolysis. (Use the graph you drew from the table on page 45 to find the water potential.)
   c   Explain the results.

9   The epidermis was peeled off some onion scale leaves, cut into pieces and placed in solutions of different concentrations of sodium chloride. The pieces were immersed for ten minutes and then placed on microscope slides in the bathing solutions. The number of plasmolysed cells in 100 cells chosen at random was counted. The results are shown below.

Concentration of sodium chloride solution/mol dm^{-3}	Percentage of plasmolysed cells
0.00	0
0.10	7
0.20	43
0.30	67
0.40	87
0.50	100

   a   Plot a graph of these results.
   b   State the water potential at which 50% of the cells are plasmolysed. (Use the graph you drew from the table on page 45 to find the water potential.)
   c   Explain how plant cells become plasmolysed.

3.1 Enzymes

Learning outcomes

On completion of this section, you should be able to:

- define the terms: *catalyst; enzyme; substrate; product; active site; activation energy*

- explain that enzymes are globular proteins that catalyse metabolic reactions

- explain that enzymes are specific to particular reactions.

∞ Link

Catabolic reactions are detailed on page 2. The other catabolic reactions you will study are those of respiration and the decomposition of hydrogen peroxide as shown on page 54.

∞ Link

Refer to page 2 to check your understanding of anabolic reactions. Other anabolic reactions you will study are those of photosynthesis.

☑ Study focus

In $n-1$ the n is a large number. Think about making a peptide out of nine amino acids, how many peptide bonds form? How many molecules of water are eliminated in the condensation reactions? Now you have the answer to the question.

☑ Study focus

A substrate molecule fits into an active site because it has a complementary shape. Remember not to write that it has the same shape as the active site.

Enzymes are catalysts

Catalysts are substances that increase the rate of a reaction without being used up themselves. **Enzymes** are biological catalysts that are made of protein. They provide a surface on which the substrate or substrates of the reaction can fit together closely and in doing so make the reaction more likely to occur. Substrates are the substances that are to be changed during the reaction. Enzymes catalyse reactions in which substrates are broken down from complex molecules to simpler ones. Examples are:

- **Sucrase**
  sucrose → glucose and fructose by breakage of a glycosidic bond
- **Amylase**
  starch → maltose by breakage of glycosidic bonds
- **Proteases**
  polypeptide → amino acids by breakage of peptide bonds
- **Lipase**
  triglycerides → fatty acids and glycerol by breakage of ester bonds
- **Nucleases**
  nucleic acids → nucleotides by breakage of phosphodiester bonds

Enzymes also build molecules by forming bonds, such as:

- glycosidic bonds between glucose monomers in starch:
  Starch synthetase
  $(glucose)_n →$ starch $+ (water)_{n-1}$
- peptide bonds between amino acids in a polypeptide:
  Peptide synthetase
  $(amino \ acids)_n →$ polypeptide $+ (water)_{n-1}$

Enzymes are specific

Figure 3.1.1 shows how enzymes catalyse reactions by providing a place for the substrate to fit. There are two ways in which this may happen. The shape of the active site is such that the substrate can fit very closely into it. This is the idea of the **lock and key**. As substrate molecules fit into the active site the enzyme molecule changes shape slightly to mould itself around the substrate so there is a better fit between the two molecules. This mode of action is known as an **induced fit**. Only enzymes with the appropriate shape of active site will accept a specific substrate. For example, amylase will accept starch, but not a protein. Amylase is specific to the breakdown of starch and catalyses the hydrolysis of every other glycosidic bond to form maltose (a disaccharide). The **specificity** of an enzyme is determined by the shape of its **active site**, which must be **complementary** to the shape of the substrate. This means that they have shapes that match or fit together.

There are different degrees of specificity and some enzymes are specific only to one reaction. Others are less specific and will catalyse a number of reactions of the same type. For example, the table shows enzymes that cut peptide bonds.

Enzyme	Specificity – breaks peptide bonds
pepsin	next to phenylalanine, glutamic acid and leucine
trypsin	next to arginine or lysine
chymotrypsin	next to phenylalanine, tyrosine and tryptophan
exopeptidases	at the C and N terminals of peptides
subtilisin	between any pair of amino acids

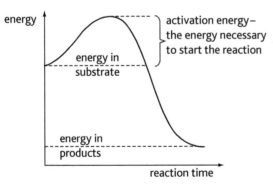

Figure 3.1.1 *The mode of action of an enzyme*

Active sites

An active site is a 'pocket' in an enzyme molecule. This is lined by R-groups of some amino acids that are close to each other after the enzyme molecule is folded into its tertiary shape. Some of the amino acids making the active site are adjacent to each other in the primary sequence, some are not. The shape made by the R-groups is important as it determines which substrate can fit. When the substrate molecule fits inside the active site the molecule is put under strain so that bonds break or form. The combination of enzyme and substrate is an **enzyme-substrate complex**.

Activation energy

The reactions catalysed by the enzymes described above are not favourable. They require the formation and breakage of covalent bonds. Covalent bonds are stable bonds, which require energy to form and break.

Figure 3.1.2 *Activation energy*

Activation energy is the energy that must be overcome before a reaction can proceed. Much energy is needed to make or break the covalent bonds in the reactions involving biological molecules. Enzymes provide active sites where substrate molecules are positioned in such a way that the activation energy is overcome. Without enzymes the reactions would occur so slowly that life would not exist as we know it.

Did you know?

There are thought to be over a thousand different enzymes in some specialised human cells.

Summary questions

1 Define the following terms: *catalyst; enzyme; substrate; product.*

2 Explain why enzymes are necessary and why there are so many in each cell.

3 a Write a word equation to show the synthesis of a triglyceride from fatty acids and glycerol.
  b State the name of the bond that forms between fatty acids and glycerol.

4 Explain the specificity of enzymes.

5 Describe the mode of action of an enzyme using the induced fit mechanism.

6 a Write out the primary sequence of a polypeptide with 20 different amino acids that includes the following adjacent pairs of amino acids: lysine and leucine, arginine and phenylalanine.
  b Show on your primary sequence the bonds that will be broken by the enzymes given in the table above.

3.2 Investigating enzyme activity

On completion of this section, you should be able to:

- describe how to follow the progress of an enzyme-catalysed reaction

- describe how to use iodine solution to follow the disappearance of starch

- explain how to calculate rates as $1/t$ for enzyme-catalysed reactions

- create tables to record results from practical investigations.

☑ *Study focus*

Keeping the enzyme solution and substrate solution separately until they are both at the desired temperature for the reaction is called **equilibration**.

The activity of enzymes is determined by finding the rate at which a reaction proceeds. When enzyme molecules are added to substrate molecules they collide to form enzyme-substrate complexes. A reaction occurs and the product molecules leave the active sites. The enzyme is now ready to form another enzyme-substrate complex. Over time the number of substrate molecules decreases and the chance of a substrate molecule entering an active site decreases. This means that the number of product molecules formed decreases.

You can follow the course of an enzyme-catalysed reaction by seeing how long it takes to reach an end point. For example a solution of milk protein is cloudy. Add a protease and the cloudiness disappears. This can be done as follows:

1 Add $10\,cm^3$ of a solution of milk powder to test-tube 1.
2 Add $1\,cm^3$ of a protease solution to test-tube 2.
3 Put both test-tubes in a water bath at $30\,°C$ for five minutes.
4 After five minutes, pour the contents of test-tube 2 into test-tube 1, return the test-tube 1 to the water bath and start a timer.
5 Watch carefully and time how long it takes for the cloudiness to disappear.

The time taken for the reaction to occur is *not* the rate of reaction. If there is a fast rate of reaction then the reaction will be completed in a short time and you can calculate the rate as: $1/t$ in which $t =$ the time taken to reach an end point in seconds.

The unit for rate in this case is seconds^{-1} (written as s^{-1}).

In many cases it is not possible to see a change such as that for the disappearance of the cloudiness as the milk protein is hydrolysed. If amylase is added to starch, there may be a slight change in cloudiness but it is better to follow the hydrolysis of starch by testing with iodine solution. To do this you take samples from the reaction mixture and test them with iodine solution and record the colour. Figure 3.2.1 shows how this is done.

☑ *Study focus*

Taking a sample as soon as the enzyme is added to the substrate is called taking a reading at 'time zero'.

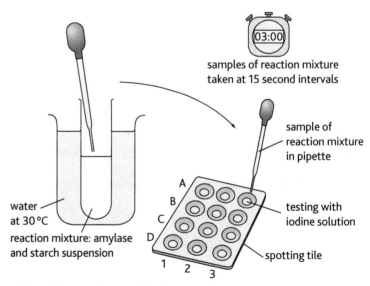

Figure 3.2.1 *Following the hydrolysis of starch*

You should continue taking samples until two results give a yellow colour indicating that all the starch has been hydrolysed. These results are recorded in a table:

Time/seconds	Colour of iodine solution	Starch present
0	blue-black	✓
15	blue	✓
30	red-brown	✓
60	yellow	✗
90	yellow	✗

When making a table to record results follow this guidance:

- make the table a reasonable size and not too small; remember not to draw it right at the top of a page
- draw the table outlines (columns and rows) in pencil
- write brief but informative headings for each column
- columns headed with physical quantities should have appropriate SI units (see page 41 for examples)
- the first column is the independent variable; the second and subsequent columns contain the dependent variable(s); sometimes you might want to put a number or letter in the first column to the left of the independent variable – for example, to identify test-tubes
- what you write in the body of the table should be brief, e.g. single words, short descriptive phrases, numbers, ticks or crosses, etc.
- data should be organised so that patterns can be seen – it is best to arrange the values of the independent variable in ascending order (i.e. down the table)
- numbers written into the body of the table do not have units (units only appear in the column headings)
- the solidus or slash (/) meaning 'per' should *not* be used in units. If you have to include concentrations in a table you should not write g per dm^3 as g/dm^3. It should always be written out in full using 'per' or, better, as $g\,dm^{-3}$. The negative exponent, cm^{-3}, means 'per'
- note that the solidus is used to separate what is measured from the unit in which it is measured. You may notice that many text books and examination papers use brackets around the units in tables.

From the results in the table we do not know exactly how long it took for all the starch to be hydrolysed. Therefore the most precise answer is 'between 30 and 60 seconds'.

Summary questions

1 Write word equations to show how the following are hydrolysed: a protein, starch, sucrose, a triglyceride.

2 What is meant by time zero? Suggest why it is important to take a result at time zero.

3 Explain why the rate for the hydrolysis of milk protein is calculated using seconds not minutes.

4 Calculate the rate of reaction from the results in the table.

5 Students often suggest that iodine solution should be added to the starch-amylase reaction mixture at time zero so that the colour change can be followed without taking samples. Suggest why this is not done.

6 Describe how you would determine the rate at which a protein is hydrolysed using the biuret test.

7 List four precautions you should take when carrying out an investigation to follow the disappearance of starch using iodine solution so that you obtain valid results.

Link

For a description of the starch–iodine complex see page 18.

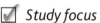

Study focus

There are 1000 milligrams in a gram. Using milligrams avoids writing numbers less than 1.0 for very low concentrations of starch. You may see micrograms (μg) used for even smaller quantities. There are 1000 μg in a mg.

In the investigation outlined in Section 3.2 the results are **qualitative**. We do not know how much starch has been hydrolysed. If we know the concentration of the solution is $10\,g\,dm^{-3}$ then $10\,cm^3$ contains 1/100 of the mass of starch = 0.1 g. So in 60 seconds 0.1 g of starch is hydrolysed and we can calculate the rate as 0.1/60 = $0.002\,g\,s^{-1}$ (grams per second) or $2\,mg\,s^{-1}$.

The rates calculated from the results in Section 3.2 are the overall rate for the complete hydrolysis of the substrate. As the substrate concentration decreases during the reaction the rate must slow down until it reaches the end point when the reaction stops. If we can take **quantitative** readings at intervals we can see if this happens.

The concentration of starch can be determined using a colorimeter to measure the optical density of the starch–iodine complex.

- Make a $10\,g\,dm^{-3}$ solution of starch and use it to make up these solutions:

  0.75, 0.50, 0.25, 0.10, 0.07, 0.05, $0.01\,g\,dm^{-3}$

- Add equal volumes of each solution to test-tubes and the same volume of iodine solution to each test-tube.
- Place each test-tube in a colorimeter to measure the optical density.

Optical density may be measured using absorbance or percentage transmission.

This table gives some colorimeter readings for known concentrations of starch.

Concentration of starch/$mg\,dm^{-3}$	Colorimeter reading: absorbance
0	0.00
1000	0.30
2000	0.52
3000	0.70
4000	0.85
5000	0.96
6000	1.06
7000	1.15
8000	1.24
9000	1.28
10 000	1.30

These results are plotted on a graph, see Figure 3.3.1.

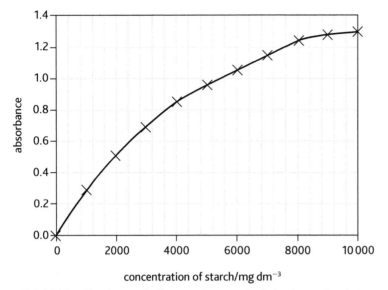

Figure 3.3.1 *This calibration graph allows you to convert colorimeter readings to concentrations of starch*

⊖⊖ Link

For advice on how to draw 'lines of best fit' on graphs refer to page 55.

☑ Study focus

Notice that the rate of disappearance of starch is not the same throughout the time period. When is it at its highest?

☑ Study focus

Use the information on page 50 to help you answer Summary question 1e.

When drawing graphs you should follow this guidance:

- make use of the graph paper so you use at least half the grid provided, do not make the graph too small
- draw the graph in pencil
- the independent variable should be plotted on the *x*-axis and the dependent variable should be plotted on the *y*-axis
- each axis should be marked with an appropriate scale, e.g. using multiples of 1, 2, 5 or 10 for each 20 mm square on the grid. This makes it easy for you to plot and extract data; never use multiples of 3
- each axis should be labelled clearly with the quantity and SI unit(s) or derived units as appropriate, e.g. time/s, concentration/g dm^{-3}, rate/g s^{-1} or rate/s^{-1}
- plotted points must be clearly marked and easy to see on the grid lines; use dots in circles (⊙) or saltire crosses (×); dots on their own should not be used; if you need to plot three lines on a graph, vertical crosses (+) can also be used.

Summary questions

1 a Copy the table of results in the margin and add a third column to show the concentration of starch.
  b Use the calibration graph to determine the concentration of starch at each sampling time.
  c Draw a graph to show the starch concentration against time.
  d Describe the trend shown in your graph.
  e Explain why the rate of reaction is not constant throughout the time period.

2 Explain the advantage of using the colorimeter when investigating the rate of hydrolysis of starch.

3 Suggest the precautions you should take when using the colorimeter to follow the disappearance of starch.

4 Describe how to make the dilutions of starch from the 10 g dm^{-3} solution.

Here are some results:

Time/min	Colorimeter reading: absorbance
0	1.32
2	0.60
4	0.30
6	0.10
8	0.09
10	0.08
12	0.06
14	0.04
16	0.01

Learning outcomes

On completion of this section, you should be able to:

- explain the term *initial rate of reaction*

- describe how to determine the initial rate of an enzyme-catalysed reaction.

If you follow the course of an enzyme-catalysed reaction you can either take samples to:

- follow the disappearance of substrate
- follow the appearance of product.

From the graph drawn on the disappearance of starch in Summary question 1c on page 53 you can observe that the rate is fastest at the beginning and decreases with time as the concentration of starch reduces. Eventually there is no substrate left so the reaction stops. The collisions between enzyme molecules and substrate molecules were highest at the beginning of the reaction as the enzyme is added to the substrate.

To follow the appearance of a product we can use the enzyme catalase, which is in yeast and is common in many animal and plant tissues. Catalase catalyses the decomposition of hydrogen peroxide.

$$2H_2O_2 \rightarrow 2H_2O + O_2$$

It is possible to use a solution of catalase, but as it is expensive it is more common to use an extract from plant material, such as potato, lettuce or celery, which also contains the enzyme. The plant material is cut up, liquidised in a blender and filtered. The filtrate is the extract and contains catalase. The diagram shows how the apparatus for following the reaction may be set up. You could also use a gas syringe to collect the oxygen produced.

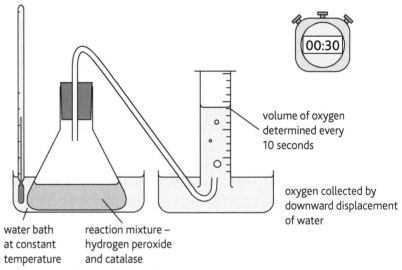

volume of oxygen determined every 10 seconds

oxygen collected by downward displacement of water

water bath at constant temperature

reaction mixture – hydrogen peroxide and catalase

Figure 3.4.1 This shows how to follow the reaction in which hydrogen peroxide is decomposed to oxygen and water

✓ Study focus

You may carry out this investigation with different apparatus. In the method shown here oxygen is collected by downward displacement of water; oxygen may also be collected in a gas syringe. The important point to note is that temperature is a control variable.

The table shows the results obtained which are then plotted on a graph.

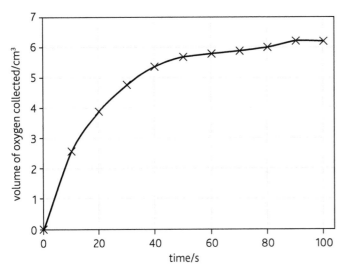

Figure 3.4.2 *Graph to show the volumes of oxygen produced when an extract of catalase is added to hydrogen peroxide*

Time/s	volume of oxygen collected/cm³
0	0.0
10	2.6
20	3.9
30	4.8
40	5.4
50	5.7
60	5.8
70	5.9
80	6.0
90	6.2
100	6.2

The initial rate is determined by either:

- calculating the rate from the first sample, or
- taking a tangent to the curve drawn on the graph.

Using the second method, the rate in this example is $0.3\,cm^3\,s^{-1}$.

Lines of best fit

After plotting the points on a graph you have to decide what sort of line you would use to show the trend. In most cases you should look to draw a line of best fit which includes all or most of the points, but this depends on what you are investigating.

You can follow these points for guidance:

- look at the points carefully – do they fall on a line of best fit? This could be a straight line that you can use a ruler to put on the graph or they could be on a curve in which case you can use a flexible ruler
- if the points do not fall neatly on a line (straight or curved), then try to place your line so you have the same number of points on either side of the line
- if you cannot see a pattern to the results, then you can join the points with straight lines which shows uncertainty about the relationship between the variables.

Catalase is one of the fastest-acting enzymes. The turnover number is the maximum number of molecules of substrate that an enzyme, such as catalase, converts to molecules of product per unit of time. Catalase has four polypeptide, each of which has an active site. Other enzymes, such as lysozyme, are smaller and only have one active site. To make fair comparisons between enzymes, the turnover number is calculated per active site of enzyme and is a measure of the efficiency of an enzyme.

Enzyme efficiency depends on the ease of fit between substrate and active site and the ability of the enzyme to catalyse the reaction. Not all collisions between active site and substrate result in a reaction. With enzymes that have high efficiency a very high proportion of such collisions are successful.

Summary questions

1 Explain why it is important to determine the initial rate of reactions in enzyme studies.

2 Describe how to determine the initial rate of reaction of decomposition of hydrogen peroxide.

3 Use the graph you drew for Summary question 1c on page 53 to determine the initial rate of reaction.

4 Predict the effect of increasing the substrate concentration on the initial rate and suggest an explanation for your answer.

5 Predict the effect of increasing the temperature on the initial rate of reaction and suggest an explanation for your answer.

Learning outcomes

On completion of this section, you should be able to:

- describe how to investigate the effect of substrate concentration on the activity of an enzyme

- explain the effect of increasing substrate concentration on the rate of an enzyme-catalysed reaction

- state the difference between competitive and non-competitive inhibitors and explain their effects

- describe how to investigate the effect of temperature on the activity of an enzyme

- explain the effect of increasing temperature on the rate of an enzyme-catalysed reaction.

Various factors affect the rate of enzyme activity. Think about our model of enzyme activity in terms of substrate molecules colliding with enzyme molecules, entering the active site where the reaction occurs and then the release of product molecules. Here we consider two factors:

- substrate concentration
- temperature.

Substrate concentration

The procedure depicted on page 54 is used to investigate the effect of substrate concentration. The procedure is repeated for each substrate concentration as follows:

- decide a range of hydrogen peroxide concentrations to be used
- decide on the intermediate concentrations to be used
- prepare the different concentrations by diluting the 20 volume hydrogen peroxide solution (20 volume means that when $1\,cm^3$ of hydrogen peroxide is decomposed, $20\,cm^3$ of oxygen are produced. A 20 volume solution is $1.67\,mol\,dm^{-3}$).

The table shows the results.

Concentration of H_2O_2/mol dm^{-3}	Volume of oxygen/cm³ time/s						
	0	10	20	30	60	90	100
0	0.0	0.0	0.0	0.0	0.0	0.0	0.0
0.2	0.0	0.7	1.4	1.9	3.2	4.4	4.6
0.4	0.0	1.3	2.1	2.8	4.0	5.2	5.5
0.6	0.0	1.8	2.6	3.2	4.6	5.7	6.0
0.8	0.0	2.1	3.2	3.9	5.2	5.9	6.0
1.0	0.0	2.0	4.0	4.8	5.8	6.0	6.0

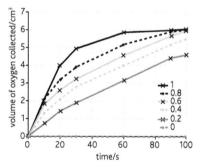

Figure 3.5.1 *This graph shows the progress of the reaction with different concentrations of hydrogen peroxide*

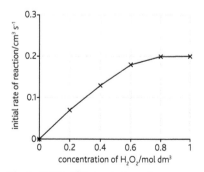

Figure 3.5.2 *This graph shows how increasing the concentration of substrate influences the initial rate of the decomposition of hydrogen peroxide by catalase*

These results are plotted on a graph (see Figure 3.5.1).

Tangents to the curves in Figure 3.5.1 are used to determine the initial rates. These are then plotted on another graph.

You can see from Figure 3.5.2 that as the substrate concentration increases, the rate of reaction increases. After $0.8\,mol\,dm^{-3}$ the rate remains constant at $0.20\,cm^3\,s^{-1}$.

At a concentration of zero there was no reaction, so the line must start at the origin. The rate increases as more substrate molecules are available and there are more successful collisions with the active sites of enzymes. Up until the concentration of $0.8\,mol\,dm^{-3}$ enzyme activity is limited by the concentration of substrate because if the concentration is increased the rate increases. At concentrations greater than $0.8\,mol\,dm^{-3}$ the rate is not limited by the substrate concentration because increasing the concentration has no effect. The rate must be limited by something else.

Some substances interact with enzymes and reduce their activity. These are known as inhibitors. Some inhibitors fit temporarily inside the active site without being changed in a reaction. These are **competitive inhibitors** because they compete with the substrate for entry to the active site. If the concentration of the substrate is increased then the effect of the inhibitor can be reduced as shown in Figure 3.5.3.

Other inhibitors combine temporarily with parts of the enzyme molecule other than the active site. These are **non-competitive inhibitors** as they do not compete with the active site, but bind to another site on the enzyme. The enzyme molecule changes its overall shape and the active site is no longer complementary to the substrate. Enzyme-substrate complexes cannot form so the enzyme is inhibited. The effect of non-competitive inhibitors cannot be reduced by increasing the substrate concentration.

Temperature

The activity of enzymes at different temperatures may be investigated by placing the reaction mixtures in water baths at a range of temperatures. The substrate and enzyme solutions are equilibrated at the temperatures used, then mixed together and kept at each of those temperatures over the range chosen (0 °C to 70 °C in this example). The results of such an investigation on protease hydrolysis of milk protein are shown in the table below.

Temperature/°C	Time for cloudiness to disappear/s	Rate of reaction × 1000/s⁻¹
0	no reaction	0.0
10	1400	0.7
15	960	1.0
20	650	1.5
25	480	2.1
30	360	2.8
35	240	4.2
40	185	5.4
50	440	2.3
55	850	1.2
60	no reaction	0.0
70	no reaction	0.0

These results are plotted on a graph (see Figure 3.5.4).

The rate of reaction increases up to a maximum rate of $5.4 \times 1000\,\text{s}^{-1}$ at the optimum temperature of 40 °C. At temperatures above the optimum the rate decreases steeply and there is no activity at 60 °C.

The rate increases as increased temperature means that there is an increase in kinetic energy and molecules of substrate collide more frequently with enzymes. However, as the temperature increases the enzyme molecules vibrate and bonds stabilising the tertiary structure begin to break. At the optimum temperature the number of successful collisions is at its maximum, but beyond that temperature more and more enzyme molecules become non-functional as they denature. At 60 °C all the enzymes are denatured.

✓ *Study focus*

Even if you did not have the result for 0 mol dm⁻³ there would be no oxygen produced so the line starts at the origin, 0,0.

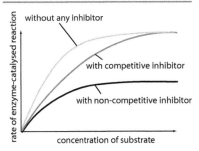

Figure 3.5.3 *This graph shows the effect of a competitive inhibitor and a non-competitive inhibitor at increasing substrate concentrations*

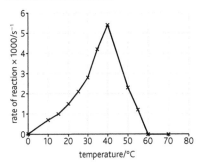

Figure 3.5.4 *This graph shows the effect of increasing temperature on the rate of an enzyme-catalysed reaction*

Summary questions

1 Explain what is meant by each of the following terms: *range, intermediate values*.

2 **a** State the limiting factor for the reaction when the substrate concentration was 0.4 mol dm⁻³. Give the evidence for your answer.

   b State the limiting factor when the substrate concentration was at its maximum in the investigation.

3 Predict the rate of reaction at the following temperatures: 35 °C and 65 °C.

4 Predict what will happen to the following reaction mixtures if transferred from: **i** 0 °C to 30 °C; **ii** 70 °C to 30 °C. In each case explain your answer.

Learning outcomes

On completion of this section, you should be able to:

- describe and explain the effect of increasing the concentration of enzyme on enzyme activity

- describe and explain the effect of pH on enzyme activity.

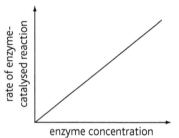

Figure 3.6.1 *This graph shows the effect of increasing enzyme concentration on the rate of an enzyme-catalysed reaction*

Enzyme concentration

From your understanding of collision theory you can probably predict the effect of increasing the concentration of enzyme on the rate of enzyme-catalysed reactions. If there are more enzyme molecules there will be an increased number of successful collisions per unit time. This can be investigated by making dilutions of the catalase extract and the initial rates of reaction can be determined as on page 54.

Figure 3.6.1 shows the results.

As the concentration of enzyme increases so the rate of reaction increases. The rate is directly proportional to the enzyme concentration. This assumes that at each concentration of enzyme the substrate concentration is in excess so that substrate concentration is not a limiting factor.

pH

Reaction mixtures may be prepared using special solutions to maintain a constant pH. These solutions are known as buffer solutions and it is possible to make buffer solutions over various ranges of pH. Three enzymes were investigated at different ranges of pH:

- pepsin
- catalase
- alkaline phosphatase.

The rates of reaction were determined and are shown in Figure 3.6.2.

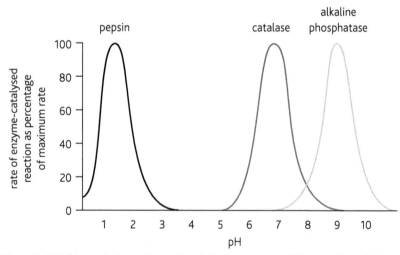

Figure 3.6.2 *This graph shows the activity of three enzymes at different values of pH*

Look carefully at Figure 3.6.2 and follow these descriptions:

- each enzyme is active over a range of pH
- as pH increases the activity of each enzyme increases until it reaches a maximum

- maximum activity occurs at the optimum pH for each enzyme
- at values of pH above the optimum pH the activity decreases
- over a certain range of pH there is no activity.

pH is the measurement of hydrogen ion concentration. At the optimum pH the enzyme is at its most active as the active site has the correct shape. The shape is determined by interactions between R-groups of the amino acids in the active site. As the hydrogen ion concentration changes some of these interactions break and the active sites become less effective. At certain values of pH the enzymes lose their shape, are no longer active and are denatured at these values of pH.

The optimum pH of an enzyme is usually close to the pH where it works. Pepsin is active in the stomach where the pH is between 1.0 and 2.0. Catalase is an intracellular enzyme and the pH of cytoplasm is about 7.0. Alkaline phosphatase is active in parts of the body with a high pH such as in bone tissue; if this enzyme enters places with a pH lower than its optimum, such as in the blood, it is not very active.

Control variables

Buffer solutions are used to give certain values of pH, e.g. pH 7.0, pH 8.5, etc. As we have seen pH is a factor that influences the activity of enzymes and it is therefore important to keep the pH constant when investigating *other* factors, such as temperature, substrate concentration and enzyme concentration. You can do this by finding a suitable pH for the enzyme that you are investigating and using a buffer solution that will keep the pH constant in each reaction mixture for as long as the reaction lasts. Variables that are not changed during an investigation but which are kept constant are called **control variables**. If you are investigating one of the factors that influences enzyme activity, then the other three must be kept constant otherwise you will not be able to draw any valid conclusions about the factor you are investigating.

Study focus

pH is not a simple linear scale. A change of one pH unit (e.g. pH 7.0 to pH 8.0) represents a change in concentration of hydrogen ions of ×10.

Study focus

You could be asked questions about enzymes with which you are familiar; you may also be asked about other enzymes which you do not recognise. Do not panic – the question will be about the principles of enzymes covered in this chapter.

Summary questions

1 Predict what will happen at higher enzyme concentrations than those shown on Figure 3.6.1 if the substrate concentration is not in excess and becomes a limiting factor.

2 Use the information in Figure 3.6.2 to complete this table.

Enzyme	Range of pH range over which enzyme is active	Optimum pH
pepsin		
catalase		
alkaline phosphatase		

3 Explain why a buffer solution is used when investigating enzyme activity.

4 Suggest how you might check that the pH does not change in a reaction mixture even though a buffer solution has been added.

5 State the control variables for an investigation into the effect of temperature on enzyme activity.

These two multiple-choice questions are of the more complex type that we introduced on page 46. Read the information very carefully. It is a good idea to write some notes to help you analyse these types of questions.

1   A metabolic pathway consists of several reactions in which the product of the first reaction is the substrate of the second and so forth. This is an example of a metabolic pathway:

$$X \xrightarrow{\text{enzyme 1}} Y \xrightarrow{\text{enzyme 2}} Z$$

What would be the effect of adding a competitive inhibitor of enzyme 2?

1   enzyme 2 would be denatured
2   substance Y would increase in concentration
3   substance Z would no longer be formed
4   enzyme 1 would be inhibited
5   substance Z would be produced at a slower rate

A 1 and 4; B 2 and 5; C 2, 3 and 5; D 5 only.   (K and U)

If a competitive inhibitor is added, then this slows down the rate at which enzyme 2 catalyses the reaction Y to Z. This will not affect the activity of enzyme 1, so 4 and 3 are incorrect. Competitive inhibitors occupy the active sites of enzymes without destroying them, so 1 is incorrect because the enzyme is not denatured. This leaves 2 and 5. If the activity of enzyme 2 is slowed down by the competitive inhibitor (see page 57), then the concentration of Y would increase as it is not used at the usual rate by enzyme 2. B is the correct answer.

2   A student prepared some test-tubes with solutions as follows:

Test-tube	Contents
1	starch and amylase
2	glucose and sucrase
3	glycogen and amylase
4	starch and sucrase
5	sucrose and amylase

The test-tubes were kept at 40 °C for 30 minutes. In which tubes would reducing sugar be detected after 30 minutes?

A 1, 4 and 5; B 1, 2 and 3; C 1, 4 and 5; D 1, 3 and 4   (K and U)

This question is about the specificity of enzymes. Amylase hydrolyses glycosidic bonds in starch and also in glycogen. (Remember glycogen is very similar to amylopectin which is a form of starch.) Although test-tube 2 has an enzyme that does not break down the substrate (glucose), there will be reducing sugar at the end because glucose is present. Sucrase is not specific for starch and amylase is not specific for sucrose. B is the correct answer.

Here are some structured questions which you can try for yourself.

3   A student investigated the effect of temperature on three protease enzymes:

P – from a prokaryote that lives in hot springs
Q – from a prokaryote that lives in the Arctic
R – from a mammal that lives in the tropics.
The results of the investigation are given in this table.

Temperature/°C	Activity of enzymes/ percentage of maximum rate		
	P	Q	R
0	0	20	0
5	0	35	10
10	0	60	25
20	8	100	35
30	15	70	48
40	30	35	95
50	45	18	35
60	65	0	0
70	100	0	0
90	40	0	0

☑ Study focus

Notice that the temperatures used do not increase evenly. Make sure you scale the axis for temperature correctly. See page 53 for guidance about this and how to plot multiple lines on a graph.

☑ Study focus

When examiners do not want to give the units in a question (possibly because they involve too much description) they will call them *arbitrary units*. You may abbreviate them to 'au' in your answers when quoting figures from your graph.

a Draw a graph of the results of the student's investigation.

b Describe the effect of temperature on enzymes, P, Q and R.

c Explain the effect of temperature on enzyme R.

d Suggest how enzymes, such as P, are able to function at such high temperatures.

4 Succinate dehydrogenase is an enzyme found inside mitochondria. It catalyses a reaction that occurs in respiration:

$$\text{succinate} \xrightarrow{\text{succinate dehydrogenase}} \text{fumarate}$$

☑ *Study focus*

Names of biochemicals

You may see succinate written as succinic acid and fumarate as fumaric acid. At the pH of the cell these biochemicals ionise to form anions (negatively charged ions with a $-COO^-$ group) not as acids with $-COOH$. They are given names with the suffix –ate to show this.

Malonate is a substance that has a molecular structure very similar to that of succinate. A student prepared a series of concentrations of succinate at pH 6.5. A small volume of malonate solution was added to each test-tube. These tubes were equilibrated at 25 °C with test-tubes of a succinate dehydrogenase solution. The enzyme solution was added to each tube and the initial rate of reaction determined. The student repeated the procedure but without using malonate. The results are shown in the table.

Concentration of succinate/arbitrary units	Rate of reaction/arbitrary units	
	With malonate	Without malonate
0	0	0
5	5	12
10	11	18
15	16	21
20	20	22
25	22	23
30	23	23
35	23	23

a Plot the results on a graph.

b Use your graph to explain the effect of malonate on the reaction catalysed by succinate dehydrogenase.

c Explain why pH and temperature are kept constant in this investigation.

5 Enzyme inhibitors may be competitive or non-competitive.

Make diagrams to show how:

i competitive inhibitors, and

ii non-competitive inhibitors interact with enzyme molecules.

6 Acetylcholine is a chemical transmitter substance released by nerve cells at synapses. It stimulates nerve cells to conduct impulses and muscle cells to contract. Acetylcholinesterase is the enzyme that breaks down acetylcholine so that it does not remain in a synapse causing continuous impulses or muscle contraction which can lead to paralysis and death.

Carbamates and organo-phosphates are insecticides that inhibit this enzyme at synapses in insects and other animals. Carbamates have a structure very similar to acetylcholine; organo-phosphates do not.

a Explain how carbamates and organo-phosphates interact with the enzyme acetylcholinesterase to inhibit its activity.

b Suggest why these insecticides are health hazards and environmental hazards if not used carefully.

6 The enzyme sucrase catalyses the hydrolysis of the glycosidic bond in sucrose.

A student investigated the effect of increasing the concentration of sucrose on the rate of activity of sucrase at 40 °C.

Ten test-tubes were set up each containing 5 cm³ of different concentrations of a sucrose solution. The test-tubes were placed in a water bath at 40 °C for ten minutes. A flask containing a sucrase solution was also put into the water bath.

After ten minutes, 1 cm³ of the sucrase solution was added to each test-tube. The reaction mixtures were kept at 40 °C for a further ten minutes. After ten minutes, the water was boiled. Benedict's solution was added to each test-tube. The time taken for a colour change was recorded and used to calculate rates of enzyme activity.

a State the independent, dependent, derived and control variables in this investigation.

b Explain why the temperature of the water bath: **i** was kept at 40 °C; **ii** why it was raised to boiling point.

c Sketch a graph to show the student's predicted results and explain the effect of increasing substrate concentration on the activity of the enzyme, sucrase.

1.1 Nucleic acids

☑ **Study focus**

Do not confuse the structure of DNA with the structure of proteins. The monomers of DNA are nucleotides; the monomers of proteins are amino acids.

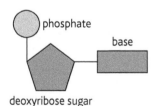

Figure 1.1.1 *A simple diagram of a nucleotide. Different shapes are used to show the five different bases.*

☑ **Study focus**

In RNA 'U replaces T'.

Remember this as you work through the next few pages. DNA has thymine, RNA has uracil.

Polynucleotides

On page 3, nucleic acids were listed as one of the four groups of biological molecules that you need to know about. There are two nucleic acids:

- **deoxyribonucleic acid (DNA)**
- **ribonucleic acid (RNA).**

These polymers are built up from monomers known as **nucleotides** which are shown in simplified form in Figure 1.1.1. DNA is a large, stable molecule that is a long-term store of genetic information. There are three types of RNA that are used for retrieving information from DNA and using it to synthesise polypeptides. The three forms of RNA are:

- **messenger RNA (mRNA)**
- **ribosomal RNA (rRNA)**
- **transfer RNA (tRNA).**

Each nucleotide is composed of a **pentose** (5-carbon) sugar, a phosphate group and one of five nitrogen-containing bases. Nucleotides are joined together to make polynucleotides by the formation of **phosphodiester bonds**. Each bond forms between the phosphate group of one nucleotide and the pentose sugar of the next.

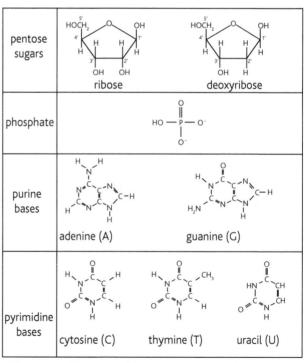

Figure 1.1.2 *The structural formulae of the components of nucleic acids*

The composition of DNA and RNA is different.

- **Ribose** is the pentose sugar in RNA – at carbon 2 there is an hydroxyl group (–OH).
- **Deoxyribose** is the pentose sugar in DNA – at carbon 2 a hydrogen replaces the hydroxyl group of ribose.
- DNA has the bases **adenine**, **guanine**, **cytosine** and **thymine**.
- RNA has the bases **adenine**, **guanine**, **cytosine** and **uracil**.

DNA

A molecule of DNA is a double helix consisting of two polynucleotides that are bonded together by hydrogen bonds. The deoxyribose sugars and phosphates make up the 'backbone' of each polynucleotide with the bases projecting inwards and forming hydrogen bonds with the bases of the opposite polynucleotide.

The nitrogenous bases are complementary in size and shape so that the only pairs that fit between the sugar–phosphate 'backbones' are A–T and C–G. The 'backbones' are arranged so that the orientation is in opposite directions – the strands are **antiparallel**. In Figure 1.1.3 you can see the sugars are arranged with carbon 3 pointing downwards on one side and upwards on the other so that the polynucleotides are arranged in a 3' to 5' direction on one side and a 5' to 3' direction on the other. Each polynucleotide assumes a helical shape as the base pairs are not at right angles to the sugar–phosphate 'backbone'.

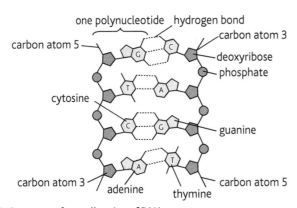

Figure 1.1.3 *Structure of a small region of DNA*

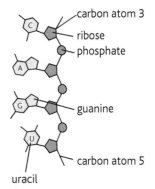

Figure 1.1.4 *Simplified diagram of a small section of RNA*

✔ *Study focus*

5' and 3' are pronounced '5 prime' and '3 prime' and are used to identify the polynucleotides in DNA and RNA. Note that one polynucleotide in DNA is 5' to 3' and the opposite is 3' to 5'.

Summary questions

1 Suggest why DNA and RNA are called nucleic acids and state the structural features that they have in common.

2 Name the five monomers of DNA and RNA and make simple labelled diagrams to show their structure.

3 State the structural differences between each of the following pairs: ribose and deoxyribose; pyrimidine and purine; DNA and RNA; polynucleotide and polypeptide.

4 Suggest why DNA is far more stable than RNA.

5 The DNA of an organism was found to contain 24% thymine. What percentages of the other three bases would you expect to find? Explain your answers.

6 Explain why the ratio of A:G and of C:T is 1:1 in DNA, but is not necessarily the same in RNA.

7 Explain in full why: **a** the ratio of A + G:C + T is always 1:1 in DNA; **b** the ratio of A + T:C + G in DNA is constant for each species but varies between species.

8 Make a table to compare the structure and functions of DNA and RNA. Do not forget to include a column for the features.

1.2 DNA replication

Learning outcomes

On completion of this section, you should be able to:

- explain the significance of base pairing in DNA
- state the significance of hydrogen bonding in DNA
- describe the process of semi-conservative replication of DNA.

Did you know?

If you pulled out all the DNA from a human cell it would extend for about 2 m. Multiply that by the number of cells in your body and it is estimated that it would extend to the sun and back 70 times!

Templates and base pairing

DNA is a store of genetic information and is passed on to new cells in growth and to new generations of organisms in asexual and sexual reproduction.

The double polynucleotide structure is ideal for **replication** because each polynucleotide acts as a **template** for making a new one. The term template means that a copy is made by matching nucleotide bases against an already existing one. Each polynucleotide has a sequence of bases and within DNA we know that adenine always pairs with thymine and cytosine always pairs with guanine. Nucleotides with these bases are assembled together along an existing polynucleotide to give a complementary sequence of bases. As this happens, the newly assembled nucleotides are joined together by phosphodiester bonds and the bases form hydrogen bonds with the template polynucleotide. This means that the nucleotides in each polynucleotide strand are held together by covalent bonds and the two polynucleotides are held together by hydrogen bonds, which collectively stabilise double-stranded DNA, but are weak enough to be broken when needed.

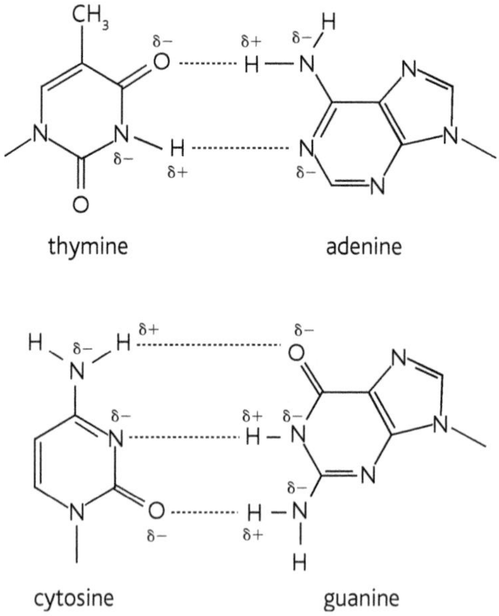

Figure 1.2.1 *There are two hydrogen bonds between adenine (A) and thymine (T) and three between cytosine (C) and guanine (G) in DNA*

Before replication can occur, nucleotides need to be synthesised. This happens in the cytoplasm and uses energy to form activated nucleotides known as deoxynucleoside triphosphates.

Replication ensures that the sequence of base pairs always remains the same, although occasionally mistakes occur to give rise to mutations (see page 128). Research by Matthew Meselsohn (1930–) and Frank Stahl (1929–) in 1958 showed that replication was semi-conservative, not conservative or dispersive as was proposed at the time. Conservative replication involved producing new DNA composed of two newly synthesised strands; dispersive replication involved making DNA composed of new sections alternating from one strand to the other.

Link

Look back to page 4 for more information on hydrogen bonds.

Study focus

Why semi-conservative? Each new DNA molecule consists of one 'old' polynucleotide that acted as the template and one newly synthesised polynucleotide. The 'old' strand has been conserved as half the new molecule of DNA – hence semi-conservative.

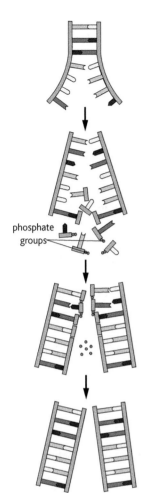

Step 1. DNA unwinds; hydrogen bonds between base pairs break. The enzyme topoisomerase unwinds the DNA and the enzyme helicase separates the two polynucleotide strands of DNA.

Step 2. Each strand acts as a template. Free nucleotides assemble against both strands.

Step 3. DNA polymerase catalyses the attachment of a nucleotide to the new, elongating strand. Two phosphates break off and a covalent bond is formed between the phosphate of the nucleotide and carbon 3 of the existing strand. DNA polymerase travels along the template strands in a 3' to 5' direction. The new strand is assembled in a 5' to 3' direction. This means that one strand is assembled in one piece, but the other is assembled in sections.

phosphate groups

Step 4. DNA polymerase checks the sequence; any incorrect base pairs are cut out and replaced.

Step 5. Hydrogen bonds form between the polynucleotide strands; topoisomerase catalyses the winding up of new DNA molecules into two double helices.

Figure 1.2.2 *Semi-conservative replication of DNA*

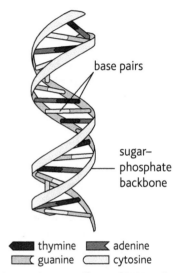

base pairs

sugar–phosphate backbone

■◤ thymine   ◤ adenine
▭ guanine   ▭ cytosine

Figure 1.2.3 *A small part of DNA to show its double helix structure*

∞ Link

Research animations of 'DNA replication' online, once you have done that try Summary question 6 and Question 2 on page 74.

Summary questions

1 Make a flow chart to show the stages in replication.

2 Use diagrams to explain what is meant by the term template in DNA replication.

3 Explain why replication is necessary.

4 State what is required in a cell before it can replicate.

5 In their paper of 1953, James Watson and Francis Crick wrote that the structure of DNA that they described 'suggests a possible copying mechanism for the genetic material'. Explain the importance of base pairing in replication.

6 Find out about the work of Matthew Meselsohn and Frank Stahl and explain how their results showed that the method of replication is semi-conservative rather than conservative or dispersive.

1.3 Protein synthesis (1)

Learning outcomes

On completion of this section, you should be able to:

- state that triplets of bases in DNA code for amino acids
- explain how the sequence of bases in DNA codes for the primary structure of a polypeptide
- use the genetic code to convert sequences of bases in DNA into sequences in RNA and into sequences of amino acids in polypeptides.

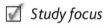 Study focus

Protein molecules have very specific shapes that are dependent on the sequence of amino acids. If the sequence changes, the protein can have a different shape and does not function properly, if at all.

✓ Study focus

Why 64 triplets? You can count the number in the table or calculate it as 4^3 ($4 \times 4 \times 4$).

✓ Study focus

There are four bases in DNA. A two base code does not work as it only codes for $4^2 = 16$ amino acids which is not enough, whereas a four base code would code for too many ($4^4 = 256$).

Overview of protein synthesis

Ribosomes assemble amino acids into polypeptides. The instructions for assembling amino acids in the correct sequence come in the form of molecules of messenger RNA which are produced in the nucleus. Some cells produce large quantities of single proteins. For example, cells in the pancreas produce insulin when the concentration of blood glucose decreases; some lymphocytes produce antibody molecules at a very fast rate. All the ribosomes in the cell require mRNA molecules that specify the sequence of amino acids for insulin or antibody. Protein synthesis involves the following:

- **transcription** of DNA to produce mRNA, which occurs in the nucleus
- activation of amino acids, which involves attaching them to tRNA molecules; this occurs in the cytoplasm
- **translation** of mRNA to form polypeptides, which occurs on ribosomes either free in the cytosol or attached to endoplasmic reticulum
- **post-translational modification** of some proteins, which occurs in the Golgi body.

Genetic code

DNA is a store of genetic information for the synthesis of polypeptides. The sequence of bases in a length of DNA is a code. Each **triplet** of bases codes for an amino acid. There are four bases in DNA (A, T, C and G) so it is possible to make 64 different triplets.

DNA is a template for making messenger RNA which conveys short-lived copies of the base sequence from the nucleus to ribosomes in the cytoplasm.

The diagram shows how the sequences of the bases in the two polynucleotides relate to the sequence in mRNA. The polynucleotide along which the sequence of RNA nucleotides is assembled is the template strand for transcription; the complementary polynucleotide is the coding strand. As you can see the coding strand has a sequence of bases which is the same as that of the mRNA produced except that U replaces T.

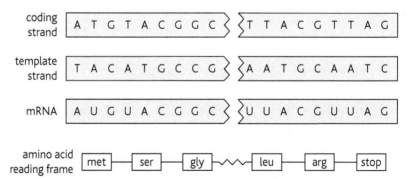

Figure 1.3.1 *Follow the sequence of bases in the two polynucleotide strands of DNA and in the single mRNA polynucleotide for the beginning and end of a polypeptide*

2nd base								
		A		**G**		**T**		**C**
1st base	**A**	AAA Phe	AGA Ser	ATA Tyr	ACA Cys			
		AAG Phe	AGG Ser	ATG Ser	ACG Cys			
		AAT Leu	AGT Ser	ATT Stop	ACT Stop			
		AAC Leu	AGC Ser	ATC Stop	ACC Trp			
	G	GAA Leu	GGA Pro	GTA His	GCA Arg			
		GAG Leu	GGG Pro	GTG His	GCG Arg			
		GAT Leu	GGT Pro	GTT Gln	GCT Arg			
		GAC Leu	GGC Pro	GTC Gln	GCC Arg			
	T	TAA Ile	TGA Thr	TTA Asn	TCA Ser			
		TAG Ile	TGG Thr	TTG Asn	TCG Ser			
		TAT Ile	TGT Thr	TTT Lys	TCT Arg			
		TAC met	TGC Thr	TTC Lys	TCC Arg			
	C	CAA Val	CGA Ala	CTA Asp	CCA Gly			
		CAG Val	CGG Ala	CTG Asp	CCG Gly			
		CAT Val	CGT Ala	CTT Glu	CCT Gly			
		CAC Val	CGC Ala	CTC Glu	CCC Gly			

2nd base								
		U		**C**		**A**		**G**
1st base	**U**	UUU Phe	UCU Ser	UAU Tyr	UGU Cys			
		UUC Phe	UCC Ser	UAC Ser	UGC Cys			
		UUA Leu	UCA Ser	UAA Stop	UGA Stop			
		UUG Leu	UCG Ser	UAG Stop	UGG Trp			
	C	CUU Leu	CCU Pro	CAU His	CGU Arg			
		CUC Leu	CCC Pro	CAC His	CGC Arg			
		CUA Leu	CCA Pro	CAA Gln	CGA Arg			
		CUG Leu	CCG Pro	CAG Gln	CGG Arg			
	A	AUU Ile	ACU Thr	AAU Asn	AGU Ser			
		AUC Ile	ACC Thr	AAC Asn	AGC Ser			
		AUA Ile	ACA Thr	AAA Lys	AGA Arg			
		AUG Met	ACG Thr	AAG Lys	AGG Arg			
	G	GUU Val	GCU Ala	GAU Asp	GGU Gly			
		GUC Val	GCC Ala	GAC Asp	GGC Gly			
		GUA Val	GCA Ala	GAA Glu	GGA Gly			
		GUG Val	GCG Ala	GAG Glu	GGG Gly			

Figure 1.3.2 These tables show the genetic dictionaries for DNA triplets and RNA codons. Amino acids are commonly known by their three-letter abbreviations.

The groups of three bases in RNA are known as **codons**. The tables above show the genetic dictionary in the form of DNA triplets on the template strand and RNA codons. The DNA triplets and RNA codons form the **genetic code** that specifies each of the amino acids during protein synthesis.

To read the DNA genetic dictionary, find the first base, e.g. A, then the second, e.g. C, and then the third, e.g. G. This triplet codes for the amino acid cysteine. Now find the complementary base sequence in the mRNA genetic dictionary. The base sequence is UGC and this is the codon for cysteine.

In reading the message for a polypeptide a cell needs a 'start' triplet as well. The triplet code for methionine (met) on the template strand is TAC and its codon is AUG. This is the start codon. Methionine is usually removed from the beginning of the polypeptide after it is produced.

Features of the genetic code

- universal – it is found in all organisms (with a few minor variations)
- it codes for the 20 amino acids found in proteins
- it codes for the sequence of amino acids that form the primary structure of polypeptides
- a sequence of three nucleotide bases codes for an amino acid
- there are three 'stop' codes
- there are between one and six codes for each amino acid (met has one and arg has six)
- there are more codes than needed so the genetic code is called a **degenerate code**
- a genetic dictionary gives the codes for the 20 amino acids plus the three stop codes
- the sequence of bases in the coding strand of DNA is the same as the mRNA (expect U replaces T)
- the sequence of bases in the template strand of DNA is complementary to that in mRNA.

☑ Study focus

You do not need to learn these triplets and codons, but expect to use a genetic dictionary and convert triplets to codons to amino acids.

☑ Study focus

Take care over your use of the terms 'triplets' and 'codons'. The groups of three bases in DNA are called triplets; groups of three bases in mRNA are known as codons.

Summary questions

1 Explain why the genetic code must be three bases and not one or two or four bases.

2 Make a table to show the DNA template codes and RNA codons for the following amino acids: phenylalanine, serine, glycine, valine, and methionine.

3 Write out the RNA sequences that code for ADH and oxytocin (see page 13).

4 Biologists talk about 'one gene one polypeptide'. Suggest what you understand by this phrase.

🔗 Link

Transcription factors bind to the region of DNA in front of the gene known as the promoter sequence, for information on this see page 123. Transcription factors 'switch on' genes during cell differentiation, for more information on this see page 76.

☑ Study focus

Termination factors bind to the termination signal region of DNA. Transcription factors and termination factors are proteins that bind to specific regions of DNA. They are examples of the specificity of protein molecules.

☑ Study focus

The free RNA nucleotides are not shown here. Each nucleotide has a base, ribose sugar and three phosphates. As in replication the loss of two of the phosphates provide the energy to form a phosphodiester bond between the nucleotide and the growing chain which becomes mRNA.

Transcription

DNA is located in the nucleus of the cells. Ribosomes that assemble amino acids to make polypeptides are situated in the cytoplasm, either free in the cytosol or attached to endoplasmic reticulum to form RER (see page 31). Each cell only has two copies of each gene, one on each homologous chromosome. It may be that only one of those genes is able to code for a functioning version of the polypeptide. The cell may need to produce large quantities of the polypeptide. The DNA cannot loop out through the nuclear pores and provide information to all the ribosomes and as a result many transcripts are produced in the form of mRNA by the process of transcription.

Transcription occurs in the nucleus of eukaryotic cells. The region of DNA to be transcribed is activated by transcription factors that identify the area of DNA to open up. Only small regions of DNA on a chromosome are activated at any one time. To access the template strand, the hydrogen bonds between the strands must break and the double helix uncoils as in replication.

The enzyme RNA polymerase is responsible for assembling the RNA nucleotides. The enzyme moves along the template strand in the 3'–5' direction. The nucleotides are assembled by forming phosphodiester bonds. The RNA polymerase stops when it reaches a sequence of bases known as the termination signal, which follows a stop triplet.

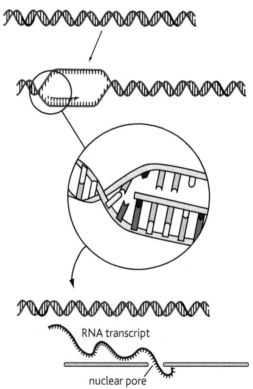

RNA transcript

nuclear pore

Figure 1.4.1 Transcription

DNA unwinds in specific areas corresponding to the gene about to be transcribed. Hydrogen bonds between the bases in the two polynucleotide chains break in the area of DNA corresponding to the gene to be transcribed. This is catalysed by helicase.

One polynucleotide acts as a template for the synthesis of mRNA. Free RNA nucleotides in the nucleus pair up with the exposed bases on the template polynucleotide.

The nucleotides are joined together by phosphodiester bonds to form a polynucleotide – mRNA. This is catalysed by the enzyme RNA polymerase (which does less proof reading than DNA polymerase).

mRNA leaves the nucleus through a nuclear pore.

The end process of transcription is an mRNA transcript. Since there are many ribosomes, the mRNA polymerase will transcribe the gene many times. These mRNA transcripts travel from the nucleus, through nuclear pores and into the cytoplasm.

Transcription is similar in many respects to replication but only occurs on one strand of DNA. Along a chromosome, some genes are transcribed from one polynucleotide, while other genes elsewhere on the chromosome are transcribed from the opposite polynucleotide.

You should know the similarities and differences between these two processes that are summarised in the table.

Features	Replication	Transcription
topoisomerase unwinds DNA	✓	✓
helicase breaks hydrogen bonds within DNA	✓	✓
number of polynucleotide strands that act as templates	2	1
free nucleotides align against DNA	✓	✓
bases of the free nucleotides	adenine, guanine, cytosine, thymine	adenine, guanine, cytosine, uracil
base pairing	A–T and C–G	A–U and C–G
name of enzyme involved	DNA polymerase	RNA polymerase
type of polynucleotide produced	DNA	mRNA
quantity of DNA involved in a nucleus	all the DNA in the nucleus	only the DNA of the genes that are active at the time

 Study focus

mRNA molecules can be used by many ribosomes at the same time. They cluster together to form polyribosomes (polysomes). mRNA molecules are broken down by enzymes and the life span of the molecules is determined as their half-life – how long it takes their concentration to decrease by a half.

 Study focus

The base pairing in transcription refers to that between the template strand in DNA and the newly formed mRNA.

 Study focus

rRNA and tRNA (see page 62) are also produced by transcription. There is some permanent base pairing in tRNA (see page 70) and in rRNA.

Summary questions

1 State what a cell needs before it can carry out transcription.

2 Draw a flow chart to show the stages of transcription.

3 Suggest why the transcription of a gene does not stop at a stop triplet.

4 Explain why RNA polymerase travels along a region of DNA many times in a short period of time, while DNA polymerase only travels along a region of DNA once per cell cycle.

5 Use Figure 1.4.1 to explain why only one polynucleotide of a gene is transcribed not both.

Did you know?

Amino acids such as lysine, which we cannot make, are known as *essential amino acids*.

⊙⊙ Link

This is a good point at which to look back to page 12 to remind yourself of the structure of an amino acid.

Transcription is the process in which mRNA is produced from a DNA template. Each molecule of mRNA is a copy of the DNA code for assembling amino acids. Once in the cytoplasm it can be translated into the primary sequence of a polypeptide. For this to happen, amino acids have first to be 'identified' or 'labelled' using the same three base codes. Since amino acid molecules do not have bases, they are attached to molecules of RNA that do. This process is **amino acid activation**. There is a pool of amino acids of all 20 types in the cytoplasm. In humans, about ten of the twenty amino acids have to be present in the diet as we do not have the ability to make them from anything else. Our cells are able to synthesise the other ten amino acids. Lysine is an example of an amino acid we cannot make and which must be in the diet.

Amino acid activation

Enzymes in the cytoplasm have active sites that accept specific amino acids and transfer RNA molecules (tRNA). The enzymes recognise the specific tRNA molecule for each type of amino acid. Energy is required for the attachment of an amino acid to its tRNA molecule. This is the only stage in protein synthesis in which the identity of the amino acid is important as after this it is identified by its tRNA molecule.

tRNA molecules have a shape resembling a clover leaf with the following regions:

- a site where amino acids are attached – always with the base sequence –CCA
- two 'loops' of nucleotides formed by some base pairing
- a 'loop' with an anticodon – the three bases that identify the amino acid.

Figure 1.5.1 shows a tRNA molecule with the anticodon UAC.

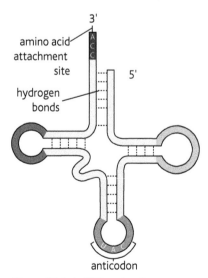

Figure 1.5.1 A tRNA molecule

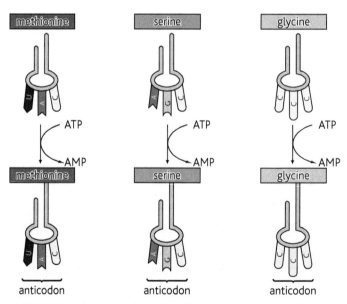

Figure 1.5.2 Amino acid activation by a type of enzyme known as an aminoacyl tRNA synthetase

The enzymes constantly activate amino acids so that there is a ready supply to take part in translation.

Translation

Now that there is a large supply of activated amino acids, the process of translation can begin. mRNA combines with the large and small sub-units of ribosomes and fits into a 'groove' between the two. Each ribosome has two sites an A site and a P site:

▪ A site accepts the tRNA-amino acid complex

▪ P site holds the lengthening polypeptide.

As you read the account below follow the translation stages in Figure 1.5.3.

To start the process a tRNA-methionine complex enters the A site. The first codon (AUG) is the start codon and only a tRNA molecule with the anticodon UAC will pair with it and therefore occupy the site. If, by chance, another t-RNA-amino acid complex enters the A site then it does not pair and does not remain in place. When the tRNA enters the A site, the ribosome moves so that the whole complex carrying methionine moves to occupy the P site exposing an empty A site. In our example, the second codon is UCG and the anticodon that pairs with it is AGC. These are the codes for serine and so a tRNA-serine complex fits into the A site. Now that both sites are full the enzyme peptidyl transferase catalyses the formation of a peptide bond between the C-terminal of methionine and the N-terminal of serine to join the two amino acids together.

Now the ribosome moves to the third codon. The first tRNA molecule leaves the ribosome and the second one carrying the dipeptide (met-ser) occupies the P site. A tRNA with the anticodon CCC occupies the now empty A site and another peptide bond forms between the dipeptide and glycine. This elongation process requires elongation factors that are on the ribosome.

This process continues until the ribosome reaches a stop codon. There is no tRNA with an anticodon for this, so translation stops and termination factors remove the mRNA molecule.

Summary questions

1   Define the following terms: *amino acid activation* and *translation*.

2   Make a table to compare tRNA and mRNA.

3   Explain the functions of the following in translation: ribosome, mRNA, tRNA and peptidyl transferase.

4   Suggest what would happen if there were no lysine in a human cell that was actively carrying out protein synthesis.

5   a   Make a list of the requirements for translation that a cell must provide.
    b   Explain the role of each item in your list.

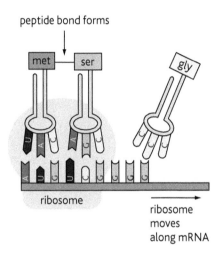

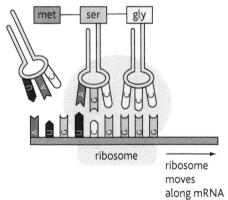

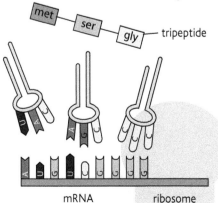

Figure 1.5.3 *Translation*

1.6 Genes to phenotypes

Learning outcomes

On completion of this section, you should be able to:

- describe the relationships between DNA, chromatin and chromosomes

- explain the terms *homologous chromosome* and *karyotype*

- discuss the links between DNA, proteins and the phenotype of an organism.

Study focus

You can find protocols for extracting DNA online by searching the term 'learn genetics'.

Figure 1.6.1 *Homologous pair of chromosomes: the human X chromosomes*

centromere

X

Link

For more details about the molecular structure of haemoglobin refer to page 16.

Chromatin

Eukaryotic cells have nuclei that you can see clearly with the light microscope if you use a stain. You can often see that the appearance of the nuclei is not even since some areas stain more darkly than others. The nuclei contain DNA that is associated with histone proteins to form chromatin. Where this is tightly coiled it takes up stain more deeply and is known as **heterochromatin**. This is inactive DNA that is not being used in transcription. The less densely staining areas are of **euchromatin** which is more active and the DNA is being transcribed.

Homologous chromosomes

Chromosomes become visible as separate structures during nuclear division. This happens after they have replicated, so the chromosomes are double structures consisting of two molecules of DNA that are packaged tightly to make two **sister chromatids**. As replication is very precise, the two sister chromatids are genetically identical.

The chromosomes are clearly visible as separate structures during the metaphase of mitosis (see page 81). If they are photographed at this stage the image can be sorted to match the chromosomes into **homologous pairs**. The arrangement of chromosomes into pairs is a karyotype. The chromosomes in each pair are the same size, they have the same shape, the centromere is always in the same place and they have the same genes. The **sex chromosomes**, X and Y, are not like this as only a small part of the Y chromosome is homologous with the X chromosome. The non-sex chromosomes are known as **autosomes**.

The human genome has about 20 000 genes and 22 pairs of autosomes and two sex chromosomes. The chromosomes are numbered 1 to 22 and two sex chromosomes. Males have 22 pairs of autosomes and an X and a Y chromosome. Females have 22 pairs of autosomes and two X chromosomes.

DNA to protein to phenotype

Genes control the sequence of amino acids in a protein. After translation, a polypeptide folds into secondary and tertiary structures. If used in the cell it will pass to the cytosol; if it is to be exported then it will pass through the rough endoplasmic reticulum to the Golgi body where it is modified, possibly by having sugar molecules added to make a glycoprotein. Modification may also involve combining with other polypeptides to form a quaternary structure or it may involve being cut and polypeptides rearranged as happens with the production of insulin.

Red blood cells form from stem cells in bone marrow. Large numbers of α and β polypeptides are formed by protein synthesis and assemble into haemoglobin molecules.

The genes for the two polypeptides are on different chromosomes. If one of the genes is faulty, then either the proper production of haemoglobin does not occur or a faulty molecule of haemoglobin is produced and the transport of oxygen is severely affected. Diseases of haemoglobin are sickle cell anaemia (see page 130) and the thalassaemias. The table shows some human genes, the polypeptides that they code for and the consequences on the phenotype if they are faulty.

Conditions in the body influence the expression of genes. For example, in the Himalayan breed of rabbit and in Siamese cats, tyrosinase is inactivated at warm temperatures and melanin is only produced in the extremities.

The phenotype of an organism is all its features, both internal and external that are the direct result of gene action (for example, sickle cell anaemia) and the interaction between genes and the environment (for example, the colour of Siamese cats).

Gene	Chromosome location	Polypeptide affected	Consequences of faulty gene
α-globin	16	α-globin	α-thalassaemia
β-globin	11	β-globin	β-thalassaemia
TYR	11	enzyme: tyrosinase (transmembrane protein in melanocytes)	albinism no melanin in the skin or hair
PAH	12	enzyme: phenylalanine hydroxylase	phenylketonuria (PKU)
AR	X	testosterone receptor	no receptor protein complete testosterone insensitivity syndrome
INSR	19	insulin receptor	no receptor protein (Donohue syndrome)

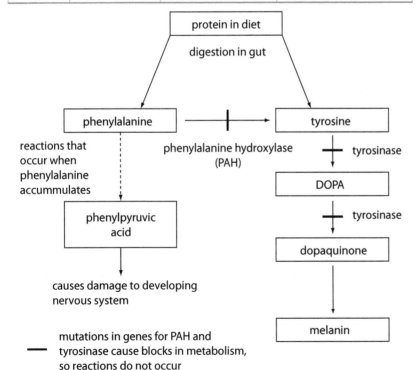

Figure 1.6.2 *The **TYR** and **PAH** genes code for enzymes that are involved in metabolising the amino acids phenylalanine and tyrosine*

∞ *Link*

Two genes given in the table code for enzymes that are involved in metabolising the amino acids, tyrosine and phenylalanine, as shown in the diagram.

Summary questions

1 Define the terms *chromatin* and *chromosome* and explain the relationship between the two.

2 Draw a flow chart to show how a named gene may influence the human phenotype. (You need to use information from this and the previous sections in this chapter.)

3 Find out about the genes that influence the following features: ability to see colour; production of factor VIII for blood clotting; testosterone receptor. What do these genes have in common?

4 Outline what may happen to protein molecules after translation.

5 Explain why insulin receptor proteins enter the Golgi body after translation, but the α and β polypeptides of haemoglobin do not.

6 Explain the effects of mutations of the genes: i *TYR*; ii *PAH*; iii *AR*; iv *INSR* on the phenotype.

All the questions in this section are short answer questions. You can find more practice MCQs on the CD that accompanies this book.

1 a Make a simple labelled diagram of a nucleotide.
   b Explain why there are four different nucleotides in DNA and not five.
   c State THREE ways in which mRNA differs from DNA.

2 In 1958, the American researchers Matthew Meselsohn and Frank Stahl published the results of their experiments on replication in bacteria. They grew bacteria for many generations in a medium containing a heavy isotope of nitrogen so that all the DNA molecules contained this heavy isotope. Some of the bacteria were transferred to a medium containing only the normal light isotope of nitrogen. After one cycle of replication all the DNA isolated from these bacteria had DNA containing both isotopes of DNA. After a further cycle of replication in the medium containing the light isotope, half of the DNA was similar to that produced in the first cycle of replication and the rest had only the light isotope.

   a Which part of a nucleotide contains the element nitrogen?
   b Explain how these results support the idea that replication is semi-conservative.
   c Predict the results that Meselsohn and Stahl would have obtained if replication was:
   i conservative; ii dispersive.

3 In the 1950s, Erwin Chargaff determined the relative quantities of the four bases in DNA in different organisms. The results, summarised in the table below, provided evidence for the structure of DNA proposed by James Watson and Francis Crick in 1953.

Organism	Percentage of adenine	Percentage of thymine	Percentage of guanine	Percentage of cytosine
Escherichia coli (bacterium)	24.7	23.6	26.0	25.7
a yeast	31.3	32.9	18.7	17.1
wheat	27.3	27.1	22.7	22.8
octopus	33.2	31.6	17.6	17.6
sea urchin	32.8	32.1	17.7	17.3
chicken	28.0	28.4	22.0	21.6
human	29.3	30.0	20.7	20.0

a Explain how the data in the table confirms the arrangement of nucleotide bases in Watson and Crick's model of the structure of DNA.

The next table shows Chargaff's data for a virus.

Organism	Percentage of adenine	Percentage of thymine	Percentage of guanine	Percentage of cytosine
a virus	24.0	31.2	23.3	21.5

b State how the data for the nucleotide bases in the virus differ from the data for the organisms in the first table.
c Suggest why the data for the virus is different from the data for all the other organisms.
d In the DNA extracted from a fish, 28% of the nucleotides contain adenine. What percentage of the nucleotides would you expect to contain cytosine? Explain your answer.
e Explain why the ratio of A + G : C + T differs from species to species.

4 Lysozyme is an enzyme found in tears and other secretions to break down the cell walls of bacteria. It is composed of one polypeptide of 130 amino acids folded into a specific tertiary structure.

The table shows:

- the sequence of three amino acids in the human lysozyme polypeptide
- one possible sequence of nucleotide bases for the mRNA that codes for these amino acids.

	arg	cys	glu
mRNA	CGU	UGC	GAA
template DNA	...	...	...

a Copy and complete the diagram to show the DNA triplets on the template strand.
b Explain why the human gene for lysozyme may have a different sequence of triplets from the answer you have given in **a**.
Lysozyme has a very specific tertiary structure.
c Explain how the DNA in the gene for lysozyme determines the specific tertiary structure of lysozyme.

5 a State THREE ways in which transcription differs from replication.

The table shows the modes of action of several drugs that inhibit replication and protein synthesis.

Drug	Mode of action
A aphidicolin	inhibits DNA polymerase
C ciprofloxacin	inhibits the enzyme topoisomerase which unwinds DNA
R rifampicin (rifampin)	inhibits RNA polymerase
T tetracycline	prevents the attachment of t-RNA to the A site of ribosomes

b Explain the effects that: **i** drugs A and C have on replication; **ii** drugs R and T have on protein synthesis.
c Drugs R and T only affect bacteria. Suggest why these drugs do not affect eukaryotic cells.

6 A length of DNA with a specific nucleotide sequence carries the code for the formation of the receptor protein for the hormone insulin. The receptor protein is a glycoprotein found in the cell surface membrane of cells in the liver, muscles and fat storage tissue.
a State the name given to the length of DNA that codes for the receptor protein.

b Outline how a sequence of nucleotides in DNA leads to the production of the membrane receptor glycoprotein for insulin.
c Suggest why the receptor for insulin is on the cell surface, but the receptor for testosterone is in the nucleus.

☑ *Study focus*

Question 6b asks for an outline, so you do not have to give all the details of transcription and translation. Also notice it says that the membrane receptor is a *glycoprotein*. You should refer to the Golgi body in your answer.

In Question 6c you need to know that testosterone is a steroid and is fat soluble.

7 The diagram shows the part of the process of protein synthesis that occurs in the cytoplasm of eukaryotic cells.

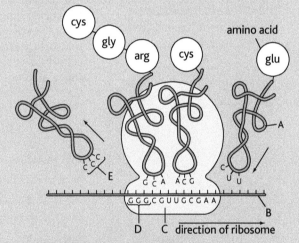

a Name the types of RNA labelled A, B and C.
b State the terms used to describe the groups of nucleotide bases at D and E.
c Explain how amino acids become arranged into the correct sequence in the primary structure of the protein. (You may refer to the diagram above to help you with your answer.)
d Describe THREE features of a polypeptide molecule that are *not* found in a DNA molecule.

☑ *Study focus*

Question 7c asks you to explain how the amino acids are arranged in the right sequence, so this answer needs details of translation.

2.1 Mitosis (1)

Learning outcomes

On completion of this section, you should be able to:

- distinguish between nuclear and cell division

- state that mitosis is the type of nuclear division that maintains genetic stability

- describe the position of replication, mitosis and cell division in the cell cycle

- discuss the role of mitosis in growth, asexual reproduction, replacement, repair and cloning.

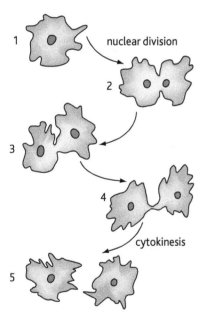

Figure 2.1.1 *This* Amoeba *is dividing into two by binary fission. The nucleus divides before the cell divides into two.*

∞ Link

Use the internet to find film of *Paramecium* dividing. Yeast reproduces asexually by budding and for further information refer to page 144.

Cell division

You started life as a fertilised egg cell, also known as a zygote. Almost immediately after fertilisation a zygote divides into two cells. This growth by cell division continues until a hollow ball of cells is formed. This embryo is about the same size as the zygote but is now composed of many cells. Each time a cell divides its nucleus divides first and then the cytoplasm divides to give two new cells, each with a nucleus. **Mitosis** is the nuclear division that occurs as cells increase in number like this. The nuclei that are produced are identical to each other and to the parent nucleus. This maintains **genetic stability**.

Later in development, the cells in the embryo gain nutrients and grow in size. When a cell reaches a certain size it is unable to support itself as the cell surface membrane is not large enough to absorb enough oxygen or lose carbon dioxide fast enough. This is because as cells become larger the surface area does not increase enough to support the increase in the volume of **protoplasm** (cytoplasm plus nucleus). As cells increase in size the surface area to volume ratio becomes smaller. This is one reason why cells divide into two after growing for a while and you can see this in unicellular eukaryotic organisms, such as *Amoeba*, *Paramecium* and yeast. Their nuclei divide by mitosis and then the cell divides by **binary fission** or by **budding**. These are forms of asexual reproduction.

Figure 2.1.2 shows the events that occur in a stem cell during one cell cycle. This shows that after a cell is formed, the following events occur in this sequence:

- growth of cytoplasm
- DNA replication
- growth of cytoplasm
- mitosis
- **cytokinesis** (cytoplasmic division).

DNA replication occurs in the 'S' phase of interphase. 'S' stands for synthesis. The 'G' phases are growth phases during which biological molecules are made to synthesise new membranes and new organelles. In G_1 molecules such as nucleoside triphosphates are made in preparation for replication and transcription. Amino acids are synthesised and attached to tRNA molecules ready for translation.

Replication almost always gives rise to identical DNA so the genetic information inherited by the two daughter cells is identical to that of the parent cell. If replication is not like this then the cells may differ genetically and not function together in a tissue. The immune system detects cells expressing different proteins and eliminates them.

Occasionally mistakes occur during replication in spite of the proof reading by DNA polymerase. Some of these influence mitosis. Differentiated cells, that have lost the power to divide, can start to divide and stem cells can lose the ability to respond to signals that control their rate of division.

In multicellular organisms, mitosis is involved in:

- growth
- repair following wounding or other damage
- replacement of cells and tissues
- asexual reproduction (see pages 144–7).

In multicellular organisms cells produced by cell division remain together to form tissues. These cells differentiate to become specialised to carry out specific functions. In animals, **stem cells** retain the ability to continue dividing by mitosis to produce more and more cells. For example, there are stem cells in the bone marrow for the replacement of red and white blood cells. Stem cells in the base of the epidermis in the skin divide to replace the cells at the surface that are rubbed off all the time. If skin is damaged by a cut then the same stem cells divide to form cells that cover the wound and then replace the damaged tissues.

In plants, **meristematic cells** are the equivalent of stem cells. Meristems are areas where these cells are found: examples are root tips, shoot tips and the cambium that gives rise to xylem and phloem tissues.

Cell cycle

The changes that occur to a cell between one division and the next are known as the **mitotic cell cycle**. During the cell cycle the nucleus divides first to form two daughter nuclei followed, in most cases, by the division of the cytoplasm to give two daughter cells each with its own nucleus. Sometimes nuclear division is not followed by cytoplasmic division as happens in fungi that have hyphae (long thin threads) that are not subdivided into cells.

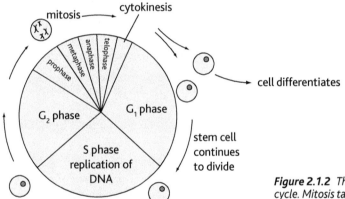

Figure 2.1.2 *The mitotic cell cycle. Mitosis takes about 5–10 per cent of the cycle.*

The rate of cell division is controlled by **proto-oncogenes** that stimulate cell division and **tumour suppressor genes** that slow cell division. A mutated proto-oncogene, called an **oncogene**, stimulates cells to divide uncontrollably. If a tumour suppressor gene mutates it becomes inactive, allowing the rate of cell division to increase. Cells that start to divide uncontrollably are often removed by the immune system; if not they may grow into tumours, such as benign and malignant cancers.

Mitosis ensures that each new cell has exactly the same genetic information as its parent cell. All the cells in the body (with the exception of gametes) have identical genetic information so that they can all work together efficiently.

As a result of mitosis:

- two daughter nuclei are produced
- the number of chromosomes in the daughter nuclei are the same as each other and the same as the parent nucleus
- genetic stability is maintained because the genetic information is the same in the daughter cells as in the parent cell
- there is no genetic variation.

There are different types of stem cell classified according to the range of cells they can form. Embryonic stem cells can produce all cell types; adult stem cells (such as bone marrow cells) produce cells of one or two tissue types.

☑ *Study focus*

Do *not* call interphase a 'resting stage'. A cell is far from being 'at rest' during interphase: DNA is replicated, macromolecules are synthesised, membranes and organelles are made.

∞ *Link*

Prokaryotes do *not* divide by mitosis since they do not have nuclei and linear chromosomes. See page 144 for a description of replication and binary fission in bacteria.

Summary questions

1 Explain the difference between *nuclear* and *cell division*.

2 Suggest and explain what would happen to a cell that does not receive a nucleus during cell division.

3 Explain why genetic stability is important during growth, repair and replacement.

4 Explain how asexual reproduction in a prokaryote differs from that in a unicellular eukaryotic organism, such as *Amoeba* or *Paramecium*.

5 Outline the events that occur during the mitotic cell cycle.

2.2 Mitosis (2)

Learning outcomes

On completion of this section, you should be able to:

- name the four stages of mitosis
- describe the processes that occur during mitosis
- make diagrams to show the four stages of mitosis.

✓ **Study focus**

Look carefully at Figure 2.2.1 and Figure 2.2.2 and identify the stages of mitosis in the photomicrograph. Then try Summary question 3.

DNA is replicated during the cell cycle and this is followed by mitosis. You can see the stages of mitosis for yourself if you make a temporary preparation of the cells from the root tip of a plant, such as onion or garlic. The growing tip of shoots and roots where mitosis is localised is called a meristem. Here is a procedure to follow with cloves of garlic.

1 Score the underside of a clove of garlic with a sharp knife and suspend the garlic over water so the base just touches the water surface.

2 After a day remove some intact roots.

3 Cut off 10–20 mm of the root tips. Put in a small volume of ethanoic acid on a watch glass (or other shallow dish) for 10 minutes.

4 Heat 10–25 cm³ of 1 mol dm⁻³ hydrochloric acid to 60 °C in a water bath. (The acid hydrolyses the DNA forming deoxyribose aldehydes which react with the stain, and hydrolyses the middle lamella that holds plant cells together.)

5 Wash the root tips in cold water for 4–5 minutes and dry on filter paper.

6 Use a mounted needle to transfer the root tips to the hot hydrochloric acid and leave for 5 minutes.

7 Wash the root tips again in cold water for 4–5 minutes and dry on filter paper.

8 Use the mounted needle to remove two root tips onto a clean microscope slide.

9 Use a scalpel to remove all but 2 mm from the growing root tip. Discard the rest, but keep the tips.

10 Add a small drop of ethano-orcein (acetic-orcein) stain or toluidine blue stain and leave for 2 minutes.

11 Break up the tissue with a mounted needle.

12 Place a cover slip over the root tips. Place filter paper over the cover slip and press gently to spread the cells. Alternatively, tap the surface of the cover slip gently with the blunt end of a mounted needle or pencil.

13 Use the low power of the microscope to search the slide for cells like those in Figure 2.2.1.

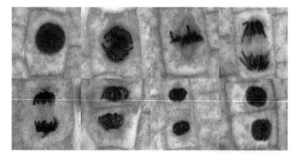

Figure 2.2.1 *Photomicrograph of plant cells dividing by mitosis*

You may be asked to make drawings from slides you have prepared in the way described above or from prepared slides. Note that the diagrams in Figure 2.2.2 are *not* drawings made from the microscope. They are diagrams to show what happens to the nucleus, centrioles and chromosomes during mitosis in an *animal cell*, not a plant cell. Plant cells do not have centrioles for organising the spindle and they have cell walls so the membrane does not form a furrow across the cell during cytokinesis. Note also that the chromosomes are drawn in a diagrammatic way – they do not look like this in real cells as you can tell by looking carefully at the photograph.

In plant cells a cell plate forms across the middle of the cell. This is made of cellulose and other cell wall materials. Membrane forms on either side of the cell plate to divide the cell into two. There is no furrow as there is with an animal cell.

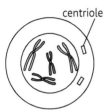
centriole

Prophase
DNA is wound up so it is packaged into chromosomes. The chromatin condenses. Each chromosome has two **sister chromatids** attached to each other at a **centromere**. Centrioles move to opposite poles of the cell. The nuclear envelope breaks up into small pieces that disperse through the cytoplasm.

Metaphase
Chromosomes arrange themselves across the middle of the cell, which is sometimes called the **equatorial plate** or **metaphase plate**. Centrioles organise the spindle apparatus by assembling **microtubules**. Some microtubules extend from pole to pole, others attach to the centromeres of the chromosomes. The DNA forming the centromeres is replicated so that the chromatids can separate in the next stage.

Anaphase
Spindle microtubules are anchored at the centrioles so when they are broken down they shorten pulling the sister chromatids apart. The sister chromatids are pulled to opposite poles with the centromere leading. Once the sister chromatids have separated they become single-stranded chromosomes, each composed of one molecule of DNA.

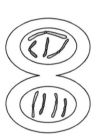

Telophase
Chromosomes arrive at the poles and uncoil. Chromatin reappears. The nuclear envelope reforms from pieces of rough endoplasmic reticulum.

Cytokinesis
During or after telophase the cytoplasm divides. In animal cells a ring of microtubules forms below the cell membrane which contracts to draw in the membrane to form a furrow. The membrane fuses so separating the two new cells.

Figure 2.2.2 The stages of mitosis in an animal cell.

 Study focus

Microtubules are made of globular protein molecules joined together to make hollow tubes. They form part of the cytoskeleton (cell skeleton) and can quickly be assembled and broken down as in mitosis.

 Study focus

Look at animations and time lapse videos of mitosis online.

 Study focus

You must be able to identify the stages of mitosis and sequence them. Remember interphase and cytokinesis are *not* stages of mitosis.

Summary questions

1 Explain why the cells in the diagrams on this page are animal cells and not plant cells.

2 Explain how mitosis maintains genetic stability.

3 Identify the stage of the mitotic cell cycle shown by each cell in Figure 2.2.1.

4 Make a drawing of one cell from each stage. (Remember to show the cell walls as two lines with a gap between them.)

5 Arrange your drawings so they show the stages as they occur in the cell cycle. (You may need to cut up your drawings and stick them into your notebook in the correct sequence.)

6 List what happens to a chromosome during a complete mitotic cell cycle.

7 Describe what happens to nuclei, nuclear envelopes, centrioles and the cell membrane during one complete cell cycle. You could tabulate your answer to make sure you describe what happens to each structure in each of the stages.

 Study focus

When answering Summary question 6, do not forget the three stages of interphase.

Did you know?

There are around 200 different specialised cells in the human body. Flowering plants by comparison have very few, but they are capable of producing a greater variety of biological molecules.

☑ Study focus

The suffix –ploid refers to the number of sets of chromosomes. You will see that cells and organisms can have one, two, three, four or more; the role of meiosis is to halve the ploidy number. Mitosis maintains the ploidy number, whatever that happens to be.

Meiosis

The other type of nuclear division is **meiosis**, in which the number of chromosomes in a nucleus is halved. This is essential in organisms that reproduce sexually so that the number of chromosomes does not double every time two gametes fuse together.

The human life cycle starts at conception when a sperm fertilises an ovum (egg cell). The fertilised egg or zygote has two sets of chromosomes – a paternal set inherited from the father and a maternal set inherited from the mother. This cell is **diploid** as it has two sets of chromosomes. All the cells produced from this diploid cell by mitosis are also diploid. But not all retain the ability to divide. They differentiate to become specialised and often lose the ability to divide. Some cells remain undifferentiated during development and enter the developing sex organs as the gamete-forming tissue. These cells are the **germ line cells**. Cells like this in the ovary divide by mitosis and then start to divide by meiosis. They stop at a time just before birth and remain in this state until after puberty when one or two start to divide approximately each month to form ova (female gametes). In the testis, germ line cells form cells that divide by meiosis to produce sperm cells (male gametes). Sperm production starts at puberty and is a continuous process making millions of sperm every day.

Meiosis is involved in the production of gametes to halve the number of chromosomes so that these cells have one set of chromosomes. They are **haploid**. Meiosis generates variation in the nuclei that are produced.

Sets of chromosomes

Diploid nuclei have two sets of chromosomes; haploid nuclei have one set of chromosomes. Each set of chromosomes contains one example of each type of chromosome. Humans have 23 chromosomes in a set with females having 22 plus an X chromosome and human males having 22 plus an X chromosome *or* a Y chromosome. The numbers of chromosomes in a set do not correspond with the size of the organism as you can see in the table. The abbreviations n and $2n$ are used to signify haploid and diploid.

Species	Haploid number (n)	Diploid number ($2n$)
fruit fly, *Drosophila melanogaster*	4	8
human, *Homo sapiens*	23	46
Pride of Barbados, *Caesalpinia pulcherrima*	12	24
Field mustard ('fast plant') *Brassica rapa*	10	20
European house mouse, *Mus musculus*	20	40
elephant, *Loxodonta africana*	28	56
domestic dog, *Canis familiaris*	39	78
adder's tongue fern, *Ophioglossum reticulatum*	630	1260*

* in common with many ferns, this species is likely to be polyploid, so there are multiple sets of chromosomes, not just two sets.

The process of halving the chromosome number is not a random process. The fact that each cell gains one complete set of chromosomes implies that it is not. Meiosis is carefully controlled so that each daughter cell gains one complete set of chromosomes with each type of chromosome represented. At fertilisation, fusion of gamete nuclei ensures that the diploid number is restored and that there are two chromosomes of each type. One set of chromosomes is maternal in origin since it came from the egg nucleus and the other set is paternal since it came from the sperm nucleus. Occasionally it is possible to tell maternal and paternal chromosomes apart, but most often not. In males the X chromosome is maternal in origin and the Y is paternal in origin.

The flowering plant life cycle is quite different to that of humans. This is hardly surprising since it is 1600 millions of years (1.6 billion years) since we shared a common ancestor. Flowers are made for sexual reproduction. They contain cells that divide by meiosis to form **spores**. The male spore is the pollen grain; the female spore is a multinucleate structure that gives rise to the female gamete and, unlike the pollen grain, is retained within the flower.

Pollen grains are transferred from flower to flower or, sometimes within the same flower. If pollen transfer is successful then a pollen grain grows a tube towards the female spore. Inside the pollen tube is a nucleus that divides by mitosis to form two male gamete nuclei. One fuses with the female gamete to form a diploid zygote and the other fuses with two other nuclei to give a **triploid endosperm** nucleus. This triploid ($3n$) nucleus then divides by mitosis to give a triploid tissue that acts as a temporary store of energy in a seed.

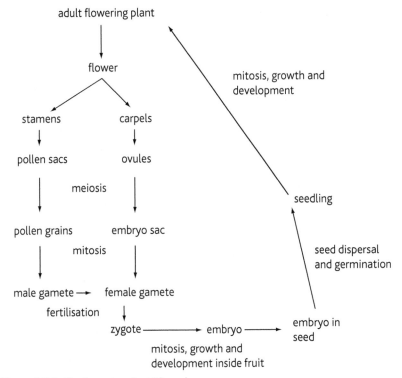

Figure 2.3.1 The flowering plant life cycle

Did you know?

Not all animals have sex determined in the same way as humans; in birds XX is male and XY is female, for example.

Link

This is an outline of the flowering plant life cycle, which is described in more detail on pages 148 to 153.

Summary questions

1   Define the terms *haploid*, *diploid* and *triploid*.

2   State the number of chromosomes in the endosperm tissue of Pride of Barbados, *Caesalpinia pulcherrima*.

3   Name ten specialised cells in humans.

4   Explain why meiosis is necessary in life cycles.

5   Make a diagram to show the life cycle of humans. Indicate where meiosis and mitosis occur on your diagram.

6   Some organisms have life cycles with short diploid stages as meiosis occurs soon after fertilisation. These organisms are haploid for most of their life cycle. One such organism is *Plasmodium* that causes malaria. Find a life cycle diagram for this organism and suggest the advantages and disadvantages of being haploid for most of the life cycle.

2.4 Meiosis

Learning outcomes

On completion of this section, you should be able to:

- state that meiosis consists of two divisions, meiosis I and meiosis II
- name the stages of meiosis I and II
- describe the processes that occur during meiosis.

☑ Study focus

Take care over writing meiosis and mitosis. It is easy to misspell them and lose marks as a result.

∞ Link

If you are unsure about what happens in mitosis and what it is for, look over pages 76–79.

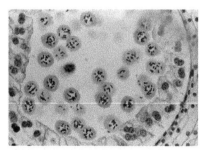

Figure 2.4.1 *Stages of meiosis during formation of pollen grains in a flowering plant*

Meiosis has many similarities with mitosis so it is easy to confuse the two. The most obvious difference is that the nucleus goes through two divisions in meiosis:

- **meiosis I** – homologous chromosomes separate
- **meiosis II** – chromatids separate.

The main ways in which meiosis differs from mitosis are as follows:

- meiosis has two divisions of the nucleus known as meiosis I and meiosis II
- *mitosis has only one division*
- meiosis I involves the pairing and separation of homologous chromosomes
- *homologous chromosomes do not associate with each other in mitosis*
- there are four daughter nuclei at the end of meiosis II
- *there are two daughter nuclei at the end of mitosis*
- the chromosome number is halved in meiosis I
- *the chromosome number remains the same in mitosis*
- daughter nuclei are haploid in meiosis
- *daughter nuclei produced in mitosis have the same chromosome number as the parent nucleus*
- daughter nuclei are genetically different from each other and from the parent nucleus in meiosis
- *daughter nuclei produced in mitosis are genetically identical to each other and to the parent nucleus*

As you look through the diagrams of meiosis I in Figure 2.4.2, follow what happens to the chromatids. You might think that if a cell divides once and the daughter nuclei have half the chromosome number that the same thing will happen again when the daughter cells divide in meiosis II. This does not happen because in meiosis II the chromatids of each chromosome separate as they do in mitosis. The chromosome number halves in meiosis I. It does *not* halve again in meiosis II.

The diagrams in Figure 2.4.2 show meiosis in an animal cell. In this case sperm are produced from the four haploid daughter cells. Details of meiosis in egg production in animals and spore production in plants are in pages 162 and 148 respectively.

The haploid cells differentiate into sperm cells (see page 162). The whole process in humans takes two months to produce mature sperm. Meiosis in potential egg cells starts before birth, but is arrested during meiosis I until the beginning of each menstrual cycle. Some eggs may remain in this arrested state for over 40 years before they complete meiosis. In the formation of an egg, the division of the nucleus is the same as shown above, but the cytoplasm divides unequally to give one large cell (the secondary oocyte) and three tiny cells that each consist of little more than a nucleus.

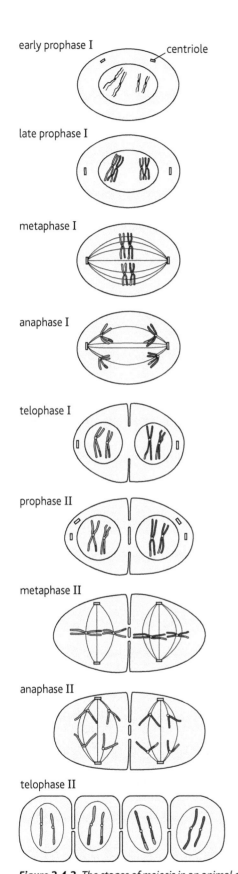

Figure 2.4.2 The stages of meiosis in an animal cell to produce sperm cells

Meiosis I
Prophase I
- chromosomes condense so that they become shorter and thicker; they are visible in the light microscope
- homologous chromosomes pair to form bivalents; in this diagram the maternal chromosomes are shaded and the paternal chromosomes are unshaded
- **chiasmata** (singular chiasma) form to hold chromosomes together; non-sister chromatids join, break and exchange parts in crossing over (see page 85 for more details); chiasmata become visible after crossing over is complete
- at the end of prophase I the nuclear envelope breaks up into small sacs of membrane, which become part of the endoplasmic reticulum; centrioles replicate and move to opposite poles and form spindle microtubules

Metaphase I
- bivalents move to the equatorial (or metaphase) plate
- the paternal and maternal chromosomes in each bivalent position themselves independently of other bivalents (in other words the paternal chromosomes are not all facing the same pole)
- microtubules attach to the centromere of each chromosome

Anaphase I
- double-stranded chromosomes (each with two chromatids) are pulled by the shortening of microtubules towards the poles; chromatids would be pulled to opposite poles if this was anaphase of mitosis

Telophase I
- chromosomes reach the opposite poles
- the nuclear envelope reforms to make two daughter nuclei that have half the number of chromosomes of the parent cell – these nuclei are haploid
- cytokinesis occurs – the cell surface membrane forms a furrow leaving small cytoplasmic bridges between the cells

Interphase
An interphase may occur between the divisions of meiosis I and II in which case the chromosomes uncoil. Cytokinesis may also occur at this stage, but not in all organisms.

Meiosis II
Prophase II
- centrioles replicate and move to the poles that are at right angles to those in meiosis I
- nuclear envelope breaks up as in prophase I

Metaphase II
- individual double-stranded chromosomes align on the equator with their chromatids randomly arranged
- microtubules attach to the centromeres

Anaphase II
- sister chromatids break apart at the centromere and move to opposite poles by shortening of microtubules as in anaphase I

Telophase II
- nuclear envelopes reform to give four haploid nuclei that are genetically different to one another and from the parent cell; cytokinesis may follow, but not in all organisms

Summary questions

1 Explain why it is important that the chromosome number:
   i halves during meiosis I;
   ii does not halve during meiosis II.

2 Describe what happens to a chromosome during meiosis I and meiosis II.

3 Make a table to compare what happens during meiosis I with what happens in meiosis II. (Remember to include similarities as well as differences.)

4 Identify the stage of meiosis shown in Figure 2.4.1 and describe what is taking place to the chromosomes.

Learning outcomes

On completion of this section, you should be able to:

- state that processes in meiosis give rise to heritable variation

- describe the process of random segregation of chromosomes in meiosis I

- explain the process of crossing over during prophase I of meiosis.

Segregation of chromosomes

All individual organisms produced by sexual reproduction are different from one another. With the exception of identical twins we are genetically unique. All members of the same species have the same genes since the types of genes determine the features of each species. Genetic differences between individuals are caused by the different versions of the **genes**, known as **alleles** that we have inherited (see page 72). The way these alleles are expressed in the phenotype depends on the interaction with the environment. Genes provide the inheritable variation. The effect of the environment on the expression of those genes is not inheritable.

During meiosis changes occur to chromosomes to rearrange the genetic information that is passed from one generation of cells to the next. The first of these changes is the halving of the chromosome number.

The chromosomes are arranged into homologous pairs so that each daughter nucleus receives one set of chromosomes with one example of each chromosome. Pairing of homologous chromosomes occurs in prophase I; separation, or segregation, of chromosomes occurs in anaphase I.

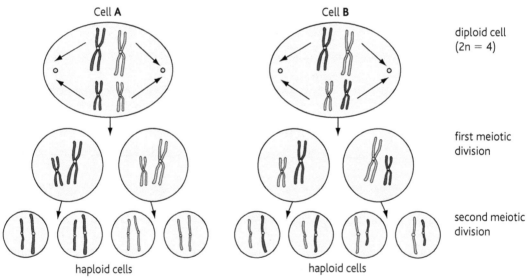

Figure 2.5.1 *Random segregation of homologous chromosomes in meiosis I. Maternal chromosomes are shaded darker; paternal chromosomes are lightly shaded.*

☑ Study focus

Make sure you know the difference between the terms *gene* and *allele*. Check the glossary for the meanings and always try to use them correctly.

∞ Link

Random segregation of chromosomes is responsible for the results of the dihybrid cross as shown on pages 98–99.

The arrangement of paternal and maternal chromosomes at metaphase I is entirely random. In Figure 2.5.1 you can see this happening with two homologous pairs of chromosomes. They can either line up as shown in cell A or as in cell B. There is a 50% chance that they will be in alignment A and a 50% chance they will be in alignment B.

This gives rise to inherited variation because each nucleus contains different mixtures of maternal and paternal chromosomes.

You can see the results of this in the haploid nuclei at the end of meiosis II in Figure 2.5.1.

Random segregation during meiosis in humans can generate 2^{23} different combinations of maternal and paternal chromosomes ($8\,388\,608$). At fertilisation the number of possible combinations is therefore $2^{23} \times 2^{23}$ in a human zygote.

Crossing over

When homologous chromosomes pair at the beginning of prophase I they wind around each other so they stay together. At each point where two chromatids touch each other the DNA breaks and sections are exchanged between non-sister chromatids. Later as the chromosomes separate it is possible to see where breakage and exchange has occurred as the chromatids remain attached to one another at chiasmata. The breakage and exchange of DNA between non-sister chromatids that occurs early in prophase I is called **crossing over**.

Crossing over gives chromosomes that are part maternal in origin and part paternal. This gives different combinations of alleles of different genes. Crossing over occurs very precisely so that genes are not gained or lost by chromatids. In fact, crossing over has to be so precise that not a single nucleotide is lost or added at the crossover point if it occurs within a gene. If this were to happen a mutation would occur and the resulting gene would produce an inactive protein or, more likely, no protein at all. Loss of nucleotides in the regions between genes may be just as harmful as there are many sections of DNA that control how genes are switched on and off.

Genes that are close together on a chromosome tend not to be split up by crossing over and are inherited together in 'clusters'. Occasionally crossing over does occur to divide them.

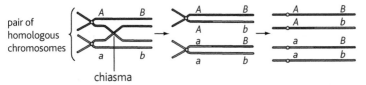

Figure 2.5.2 *Crossing over between chromatids of a homologous pair and the effect on the alleles of two genes,* **A/a** *and* **B/b**

Crossing over adds even more variation to that generated by random segregation of chromosomes. As a result of all this variation it is thought that no two humans, with the exception of identical twins, have ever had or ever will have the same genotype.

Meiosis promotes variation

The effects of random segregation and crossing over are to 'shuffle' the genes that an individual has inherited before passing them on to the next generation. We have seen four ways in which meiosis generates variation that is inheritable:

- the alleles of one gene segregate so daughter nuclei contain one of a pair of alleles
- homologous chromosomes show random segregation so daughter nuclei contain a mixture of paternal and maternal chromosomes
- homologous chromosomes cross over during prophase I so that daughter nuclei receive chromosomes that are a mixture of paternal and maternal segments
- the segregation of sister chromatids is also random – often one chromatid is a mixture of maternal and paternal DNA while the other one is not.

If meiosis I halves the chromosome number, why is there a need for a second division? At the end of meiosis I, each chromosome has two sister chromatids. In order for replication to occur these chromatids will have to separate to restore the diploid number. Also crossing over occurs so that the sister chromatids are not identical. If a chromosome with two chromatids is in a gamete and inherited by the next generation there will be two groups of cells in the body that are not genetically identical and will not work together.

Summary questions

1. Explain what is meant by the term *heritable variation*.

2. Define the term *segregation* as it is used to describe the behaviour of chromosomes.

3. a State when random segregation of chromosomes occurs during meiosis, and
   b describe the process of random segregation.

4. Explain why chiasmata are formed during prophase I.

5. a State when crossing over between chromatids occurs, and
   b describe the process of crossing over.
   c Make some diagrams to show the effect of crossing over on the alleles of three genes: **A/a**; **B/b**; **C/c**.

6. Explain the advantages of meiosis in a life cycle.

7. Explain why identical twins have the same genotype.

On completion of this section, you should be able to:

- use and make tables to compare mitosis and meiosis
- identify the events that occur during each stage of meiosis
- make models of chromosomes to show what happens during mitosis and meiosis.

☑ *Study focus*

Think about each feature in this table and explain the importance of each in nuclear division.

This chapter concerns two forms of nuclear division, mitosis and meiosis.

It is a good idea to summarise the similarities and differences between mitosis and meiosis in a table.

Feature	Mitosis	Meiosis
during the process the nuclear envelope 'disappears'	✓	✓
DNA condenses at the start of division	✓	✓
number of divisions of the nucleus	1	2 (meiosis I and meiosis II)
The chromosome number of the daughter nuclei	same as the parent nucleus	half the number of the parent nucleus
homologous chromosomes pair to form bivalents	✗	✓
crossing over occurs	✗	✓ in meiosis I
chromosomes are arranged at the metaphase plate	✓	✓ in both divisions
random segregation occurs	✗	✓ in meiosis I
centromeres divide	✓	✓ in meiosis II only
chromatids move to opposite poles of the cell	✓	✓ in meiosis II only
genetic variation among daughter nuclei	✗ (maybe a little if errors in replication)	✓ (promoted by crossing over, random segregation of chromosomes)

Compare the table above with the table you drew to compare meiosis I and meiosis II.

1   Make a new table that compares mitosis with meiosis I and meiosis II. You should be able to think of other features to add to the table (see Question 3 on page 83).

The table below is a list of stages of meiosis (A to H) and a list of events that occur during meiosis (1 to 10).

2   Match each event with a stage or stages of meiosis. Use the numbers and letters to identify the events and stages; you may use each letter once, more than once or not at all. Some events occur in more than one stage.

Event during meiosis	Number
pairing of chromosomes	1
condensing of chromosomes	2
nuclear envelope reforms	3
bivalents align on equator of cell	4
crossing over	5
centromeres divide	6
separation of homologous chromosomes	7
separation of sister chromatids	8
chromosomes align on the equator of the cell	9
four nuclei are formed	10

Stage of meiosis	Letter
prophase I	A
metaphase I	B
anaphase I	C
telophase I	D
prophase II	E
metaphase II	F
anaphase II	G
telophase II	H

✔ *Study focus*

Devise your own matching question for mitosis.

3   The DNA content of cells changes throughout the cell cycle. This graph shows the changes that occur.

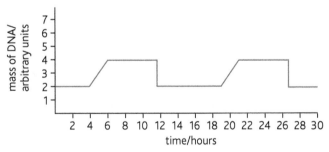

Figure 2.6.1 *Changes in DNA content in a cell during a mitotic cell cycle*

Explain what happens at each stage of the cell cycle to bring about changes in the quantity of DNA shown in Figure 2.6.1.

4   Make models of chromosomes using pipe cleaners or other suitable material such as knitting wool, string or modelling clay. Use them to model the different stages of mitosis and meiosis.

Explain the advantages of making models to show the behaviour of chromosomes during nuclear division. What are the limitations of these models?

✔ *Study focus*

Print out separate photos of the stages of mitosis. Write notes on the events that occur in each stage of mitosis on separate cards. When you revise match the photos with the notes.

This section contains SAQs on mitosis and meiosis. You can find MCQs on the CD at the back of this book.

1 The diagram shows a mitotic cell cycle.

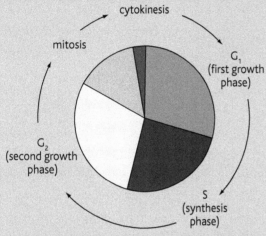

a Outline the processes that must happen during interphase before a cell divides by mitosis.

b Name the stage of mitosis in which each of the following occurs:

i chromosomes split at centromeres; ii chromosomes condense; iii sister chromatids move to opposite poles; iv nuclear envelope reforms; v chromosomes assemble at the equator of the cell.

c Explain the role of mitosis in the growth of animals and plants.

d State THREE other roles of mitosis in animals and plants.

e Suggest what might happen to the daughter cells at the end of one mitotic cell cycle.

2 The two photographs show stages of mitosis.

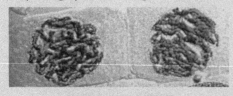

a Describe what happens to the chromosomes BETWEEN the two stages of mitosis shown in the photographs.

The next photograph is an electron micrograph of a lymphocyte in another stage of mitosis.

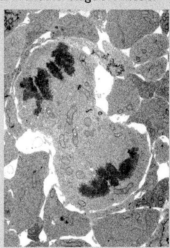

b Describe what is happening in the lymphocyte during the stage shown in the electron micrograph.

c Explain the advantages of using a light microscope to study the behaviour of chromosomes rather than the electron microscope.

☑ *Study focus*

Look at some time lapse film of mitosis before answering Question 2c.

3 A student made a series of sections through the root tip meristem of an onion. The student counted the number of cells in each stage of the cell cycle and presented the results in a table. The student discovered that it takes 13 hours at 25 °C to complete the cell cycle.

Stage of cell cycle	Number of cells	Percentage of cells in each stage	Length of time of each stage/min
interphase	254		
prophase	45		
metaphase	16		
anaphase	7		
telophase	23		
total	345	100	780

a Copy and complete the table by: **i** calculating the number of cells in each stage as a percentage of the total number counted; **ii** calculating the length of time of each stage.

b Suggest the conclusions that the student could make about the relative length of time of each stage in the cell cycle.

c Describe what happens to a root tip cell during and immediately after telophase.

4 a At the end of interphase of the mitotic cell cycle, human cells have 6 picograms of DNA in each nucleus. The diploid number for humans is 46. Copy and complete the table to show how much DNA and how many chromosomes there are in human nuclei at different stages of mitosis and meiosis.

Stage of nuclear division	Mass of DNA per nucleus/ pg	Number of chromosomes per nucleus
prophase of mitosis		
end of telophase of mitosis		
prophase I of meiosis		
prophase II of meiosis		
end of telophase II of meiosis		

b Explain how mitosis maintains genetic stability during the growth of an animal.

c Explain how meiosis and fertilisation give rise to variation within a population of animals.

5 a Explain what is meant by the terms *haploid* and *diploid*.

The Indian muntjac deer, *Muntiacus muntjak*, has the lowest diploid number of any mammal. Female muntjac deer have a diploid number of six.

b Draw diagrams to compare the arrangement of chromosomes in metaphase of mitosis with the arrangement at anaphase I of meiosis in this species.

c Describe THREE ways in which the behaviour of chromosomes during meiosis differs from that of chromosomes in mitosis.

6 The diagram shows the life cycle of *Chlamydomonas reinhardtii*, a single-celled organism. The adult organisms divide to form gametes that are of different mating strains: + and –. The haploid number of this species is 17.

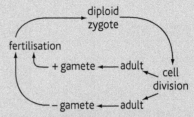

a Copy the life cycle diagram and indicate where meiosis and mitosis occur.

b State the number of chromosomes in the zygote and in the gametes.

c Explain why all the gametes produced from one adult organism are genetically identical to each other.

d Describe how the life cycle of this single-celled organism differs from that of humans.

7 The human diploid number is 46. Human chromosome No. 1 is the largest and, apart from the Y chromosome, No. 22 is the smallest.

a Make a diagram to show how random segregation of chromosomes No. 1 and No. 22 at anaphase I of meiosis leads to variation.

b i Make a diagram to show crossing over between non-sister chromatids of chromosome No. 1.

   ii Explain how crossing over increases variation.

c Describe what happens to chromosomes No. 1 and No. 22 during anaphase I and anaphase II of meiosis.

8 a Make a diagram of ONE pair of homologous chromosomes as they would appear at metaphase I in meiosis.

b Use your diagram to explain what is meant by the term homologous when applied to chromosomes.

c i State how the chromosomes you have drawn behave differently during mitosis;

   ii Explain why the behaviour of chromosomes in meiosis is different to that in mitosis.

3.1   Introduction to genetics

Learning outcomes

On completion of this section, you should be able to:

■ define the terms *inheritance*, *genetics*, *phenotype*, *genotype*, *gene* and *allele*

■ explain the relationship between the genotype and phenotype of an organism.

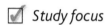 Study focus

Find out how the phenotypic features listed here are controlled and inherited.

Did you know?

Mendel was lucky in his choice of plant and features to study. This was because he had observed that pea plants tend to self-pollinate but can be cross-pollinated by hand and the features he chose are inherited in a discrete (either/or) fashion.

Study focus

It is possible to repeat Mendel's experiments using specially bred 'fast plants' that complete their life cycle within 35 days. The seeds can be harvested and then germinated immediately so it is possible to get results much faster than relying on plants that have one generation per year.

Human inheritance

We all know that we inherit features from our parents and pass on such features to our children. People sometimes say that certain human features 'skip a generation' from grandparents to their grandchildren. **Inheritance** is the passing of features from one generation to another.

Here is a list of some human features that are inherited to a greater or lesser extent.

- hair colour, length and style
- skin and eye colour
- free or attached ear lobes
- intelligence
- blood groups, e.g. ABO, MN and Rhesus
- height

- brachydactyly (short fingers and toes)
- polydactyly (six fingers and toes)
- freckles on the face
- hitchhiker's thumb
- ability to taste phenylthiocarbamide (PTC)
- hand span

All the features listed above are part of the human **phenotype**. Some are controlled only by the genes that we have inherited to give our **genotype**; the environment has no effect. Some are controlled by one gene (monogenic) and some by many genes (polygenic); many human features are influenced by both genes and our environment.

Phenotype refers to all the features of an organism other than its genes. These features are anything to do with biochemistry, physiology, morphology (outward appearance) and anatomy.

Genotype is the genetic constitution of an organism; it usually refers to the alleles of one or two genes as you will see in the following pages.

In the mid-19th century, Gregor Mendel (1822–84) carried out a systematic study of the inheritance of different features in the garden pea, *Pisum sativum*.

At the time no-one understood how features were inherited and many assumed it involved some sort of blending from both parents. Mendel was the first to state that features are controlled by discrete entities that are passed on from parent to offspring. He deduced the particulate nature of inheritance involving **genes**. We now define a gene as a length of DNA on a chromosome that codes for the production of a polypeptide. The nucleotide sequence can vary to give different forms of the polypeptide that function differently. These different forms are known as alleles.

Mendel published his results in 1866 in a scientific journal without a wide readership. His work was not appreciated until his paper was rediscovered at the beginning of the 20th century. In 1905, William Bateson (1861–1926) coined the term **genetics** to mean the study of inheritance and study of variation. Today the term covers all the aspects of genes, their effects, methods of control, inheritance and DNA technology, such as genetic engineering and gene therapy.

Mendel had no idea about the mechanism of inheritance. Chromosomes were discovered in the 1870s and it was not until 1952 that DNA was confirmed as the material of inheritance.

Animal and plant breeding

Barbadian blackbelly sheep are descended from sheep brought to Barbados in the 17th and 18th centuries. Farmers selected sheep to breed from and they chose the most suitable offspring to continue to breed animals with features adapted to conditions on the island.

The breeding of Jamaica Hope cattle began in the early years of the 20th century at Hope Farm to produce a breed that was adapted to tropical conditions. Two European breeds, Jersey and Holstein, give high yields of milk but are not adapted to tropical conditions. African Zebu cattle were crossed with the European breeds to introduce the features of disease resistance and hardiness to tropical conditions. The genotype of the current Jamaica Hope breed is about 80% Jersey, 15% Zebu and 5% Holstein.

Figure 3.1.1 *Barbadian blackbelly sheep*

Features of Barbadian blackbelly sheep:

- heat tolerant
- grow coarse hair instead of wool (so avoid overheating in the tropics)
- disease and parasite resistant
- breed throughout the year (unlike other breeds of sheep)
- give lean and mild-flavoured meat even on poor food
- have more stamina and are more agile than most breeds of sheep.

Features of Jamaica Hope cattle:

- heat tolerant
- good resistance to ticks (insect external parasites) and tick-borne diseases
- high milk yield
- survive on poor pasture.

Many of the features that breeders try to improve, such as milk yield and carcass weight, are quantitative features and are controlled by many genes and are influenced by the environment. This makes improvement difficult as so many factors are involved. Qualitative features, such as presence or absence of horns in cattle, are controlled by single genes and improvement may take only a few generations. These features are more important to breeders of named breeds. Commercial farmers are more interested in quantitative features.

Some plant diseases, such as wheat rust and black Sigatoka that attacks bananas, threaten to wipe out crop plants. We need crops that have resistance to these diseases; we also need crops that grow in saline soils and tolerate arid conditions to cope with climate change and the degradation of land that has happened as a result of irrigation and poor farming practices. We can only secure the future of our food supply by understanding the genetics of our livestock animals and crop plants.

Summary questions

1 Explain what is meant by the following terms: *inheritance, genetics, phenotype, genotype, gene* and *allele*.

2 Explain how Barbadian blackbelly sheep and Jamaica Hope cattle are adapted to tropical conditions.

3 Explain: **i** the terms qualitative features and quantitative features as they apply to genetics; **ii** why it is difficult to breed cattle to improve their milk yield.

3.2 Terminology in genetics

Learning outcomes

On completion of this section, you should be able to:

- state that a gene is a length of DNA that codes for the production of a polypeptide
- state that the locus is the position of a gene on a chromosome
- make genetic diagrams to explain the results of crosses and make predictions.

Link

During meiosis homologous chromosomes separate, so that each haploid nucleus receives one allele of each gene. See page 84 for more information on this.

Did you know?

The term *hybrid* is used for the offspring of crosses between different varieties of the same species and for crosses between different species.

Did you know?

Chromosomes are stained to show the characteristic banding patterns that you can see in this drawing of an X chromosome. Each band represents a region of the chromosome containing many genes.

Genes and chromosomes

In a differentiated cell each chromosome is single-stranded and so composed of one DNA molecule. Genes are the sections of this DNA that are used by cells to code for polypeptides. In between these **structural genes** there are other lengths of DNA that do not code for polypeptides, but control when transcription is turned on and off. There are lengths of DNA, known as **promoter sequences** to which RNA polymerase binds. In between genes there are areas of non-coding DNA that have a variety of uses. Some of it may be redundant.

It is possible to pinpoint the position of each gene on a chromosome. The position of a gene on a chromosome is its **locus** (plural **loci**). Figure 3.2.1 shows the loci of some of the genes on a human X chromosome.

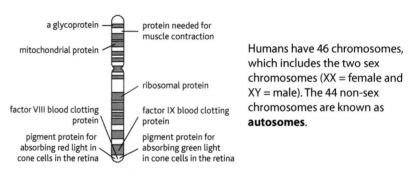

Humans have 46 chromosomes, which includes the two sex chromosomes (XX = female and XY = male). The 44 non-sex chromosomes are known as **autosomes**.

Figure 3.2.1 *A human X chromosome showing some of the gene loci*

Mendel's laws of genetics

Mendel established pure bred lines of pea plants that always showed the same characteristic generation after generation. As pea plants self-pollinate they can be **inbred**. He crossed different varieties of pea plants showing alternative features, e.g. tall and dwarf to produce **hybrid** offspring. All the plants in this generation were tall. When these plants were self-pollinated the next generation consisted of tall plants and dwarf plants in a ratio of 3:1. We explain Mendel's results by stating that there is a gene that controls height and it has two alleles – tall and dwarf. Each individual has two alleles for this gene. In order to give these results the allele for tall must be dominant over the allele for dwarf. Alleles retain their identity during gamete formation. There is no 'blending' inheritance. The alleles separate during meiosis to produce the gametes of the hybrid generation. Mendel called this his first law – the law of **segregation of alleles**. This segregation is the result of the separation of homologous chromosome during anaphase I of meiosis (see page 84).

When Mendel investigated the inheritance of two genes he discovered that each gene behaved in the same way as when he looked at the inheritance of one gene only. For example, when tall pea plants with purple flowers were crossed with dwarf plants with white flowers; all of the next generation were tall plants with purple flowers. When these plants were self-pollinated the next generation consisted of all four possible phenotypes in the ratio 9:3:3:1, which as you will see on page 99 means that the allelic pairs (tall + dwarf and purple + white) segregated independently from one another. Mendel called this his second law – the law of **independent assortment**.

Figure labels: a glycoprotein; protein needed for muscle contraction; mitochondrial protein; ribosomal protein; factor VIII blood clotting protein; factor IX blood clotting protein; pigment protein for absorbing red light in cone cells in the retina; pigment protein for absorbing green light in cone cells in the retina

Genetic diagrams

Genetic diagrams are used to show how the genes for different features are inherited. These should be set out clearly so that teachers and examiners can follow your reasoning; you should not take any short cuts. The rules for making genetic diagrams are explained below.

Step 1: Describe the gene or genes concerned.

Each gene controls a certain feature. Make sure that you state this if it is not given in the question. Mendel investigated genes that control plant height, position of flowers on the stem and flower colour.

Step 2: Identify the allelic features.

There will always be at least two alleles for each gene. The information provided will tell you which allele is dominant or it will be obvious from the description of the genetic cross.

Step 3: Choose symbols for the alleles.

Use a capital letter for the dominant allele and a lower case letter for the recessive allele. Use the same letter, not two different letters. If there is codominance or multiple alleles, then use a capital letter for the gene and superscripts for the alleles (see pages 96 and 102).

Step 4: Give the phenotype and genotype of the parents.

The genotype consists of the alleles of the gene concerned. Each cell is diploid and has two copies of each chromosome. The genotype reflects this by consisting of two alleles, even if they are the same. Some genetic diagrams start with two homozygous individuals, but do not expect all of them to do this. Test crosses may involve crossing a heterozygous individual with a homozygous recessive individual.

Step 5: State the genotype of the gametes.

Gametes are haploid because of meiosis halving the chromosome number. Always write the genotypes of gametes inside circles. In monohybrid crosses there will be one letter, in dihybrid crosses there will be two in each circle. Write down the different types of gamete. If they are all the same just write down one genotype in a circle.

Step 6: Use a Punnett square to show all the possible genotypes that result from fusion of gametes at fertilisation. This square gives the probability of each genotype, not the actual numbers of offspring.

Step 7: Write out the different genotypes and give the phenotype of each one.

Step 8: Write out the probability of each phenotype and express the answer as a ratio.

Sometimes you will be given the numbers of offspring. Note that these are numbers for different categories, so called categoric data. Always divide numbers by the smallest number to find the overall phenotypic ratio.

Summary questions

1 Define the following terms: *autosome*, *sex chromosome*, *haploid* and *diploid*.

2 Explain what is meant by the following: *inbreeding*, *interbreeding*, *cross breeding* and *hybridisation*.

3 Explain the terms *gene* and *locus*.

☑ *Study focus*

Follow these steps carefully when answering questions using genetic diagrams in this study guide and in your examination.

∞ *Link*

Look at the definitions of phenotype and genotype on page 90. Here they are used in a narrow sense to refer to the one or two features that we are studying and the gene or genes that control them.

∞ *Link*

If you are writing a genetic diagram that shows sex linkage as on page 97 then include the X and Y chromosomes in the genotypes.

☑ *Study focus*

Do not use criss-cross lines to show the possible genotypes as it is easy to make mistakes

∞ *Link*

You will see on pages 104 to 107 that a statistical test is applied to categoric data.

On completion of this section, you should be able to:

- state that a monohybrid cross involves the inheritance of one gene
- define the terms *dominant*, *recessive*, *homozygous* and *heterozygous*
- make genetic diagrams to interpret monohybrid crosses.

☑ *Study focus*

The stages in the life cycle of a fruit fly are egg, larva (maggot), pupa and adult. Metamorphosis occurs during the pupal stage as the larva changes to an adult.

☑ *Study focus*

Take care over choosing letters to represent genes. You should choose the first letter of the dominant feature unless the capital letter looks too similar to the lower case letter in your handwriting, which might cause confusion. If this is the case choose another letter.

🔗 *Link*

Look back to the advice on making genetic diagrams on page 93.

Confirming Mendel's ratios

The fruit fly, *Drosophila melanogaster*, was the animal chosen by Thomas Hunt Morgan (1866–1945) for studying genetics early in the 20th century in New York. Morgan originally set out to disprove Mendel's theories about inheritance, but in fact he found that the laws of segregation and independent assortment applied just as much to fruit flies as to peas. He also provided evidence for the theory that genes are located on chromosomes by investigating sex linkage (see page 97). There are many advantages to using *Drosophila*, but one big disadvantage – it flies! In spite of this they are still bred for use in schools and colleges and are used in research into genetics and development.

In a population of fruit flies some emerge from pupae with very small wings. In the wild these do not survive, but in captivity they do.

Pure-bred lines of fruit flies with small wings and long wings were established. When male and female flies from these pure-bred lines were crossed the offspring all had long wings. When these fruit flies were bred together three quarters of them had long wings and a quarter had short wings.

This feature is controlled by a gene with a length of DNA coding for a protein that influences the development of the wings. The two forms of the gene, long and short, are known as alleles.

The genetic diagram below shows a monohybrid cross involving the inheritance of one gene – the gene for wing length in *D. melanogaster*. There are two alleles: long wing (also known as wild type) (**W**) and vestigial (**w**). Vestigial means very small.

parental phenotypes	male long wing (wild type)	×	female vestigial wing
parental genotypes	**WW**		**ww**
parental gametes	(**W**)	+	(**w**)
F_1 *genotype*		**Ww**	
F_1 *phenotype*		all long wing (wild type)	
F_1 *phenotypes*	long wing	×	long wing
F_1 *genotypes*	**Ww**		**Ww**
F_1 *gametes*	(**W**),(**w**)	+	(**W**),(**w**)

		male gametes	
		(**W**)	(**w**)
female gametes	(**W**)	**WW**	**Ww**
	(**w**)	**Ww**	**ww**

F_2 *genotypes and phenotypes*	**WW** long wing	**2Ww** long wing	**ww** vestigial wing
F_2 *phenotypic ratio*	3 long wing : 1 vestigial wing		

The checkerboard or Punnett square is the best way to show the genotypes of the next generation – even when there are few genotypes involved. Note that the Punnett square shows the possible outcomes – the genotypes do not represent actual organisms. It shows the probabilities of different genotypes and phenotypes in the next generation.

During meiosis in the F_1 flies, the alleles separate because they are on homologous chromosomes that separate during meiosis I. This separation is **segregation of a pair of alleles**. Notice also that fertilisation gives rise to variation since gametes with the allele **W** can fuse with gametes with **W** or **w** to give flies with the homozygous dominant genotype (**WW**), flies which are heterozygous (**Ww**) and flies which are homozygous recessive (**ww**).

If the cross is carried out by using females with long wings and males with short wings (the reciprocal cross) the same results are obtained. This shows that the gene for wing length is not linked with the inheritance of the chromosomes that determine sex.

Monohybrid test cross

Once we know which allele is recessive, we know that all individuals with the recessive characteristic are homozygous recessive. However, the genotype of an individual with the dominant characteristic could be homozygous dominant or heterozygous. We can find out by doing a **test cross**. This involves crossing the individual with the unknown genotype with an individual that is homozygous recessive.

A male fruit fly with long wings is crossed with a female that is homozygous recessive. The genetic diagram shows how to work out the genotype of the male.

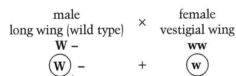

	male	×	female
phenotypes	long wing (wild type)		vestigial wing
genotypes	**W –**		**ww**
gametes	Ⓦ , –	+	Ⓦ

		male gametes	
		Ⓦ	⊝
female gametes	Ⓦ	**Ww**	**– w**

	if – = **w**	*if* – = **W**
offspring genotypes	**Ww ww**	**Ww Ww**
offspring phenotypes	50% long wing : 50% short wing	all long wing (wild type)

explanation

If all the offspring have long wings, then the gametes that have been given as ⊝ must have the dominant allele, **W**.

If some of the offspring have short-wings, then the gametes given as ⊝ must have the recessive allele, **w**.

In fact, if any short winged flies emerge we know that the male must have passed on recessive alleles and he must be heterozygous.

☑ *Study focus*

The terms F_1 and F_2 stand for first filial and second filial. They should only be used when the parental individuals are homozygous.

☑ *Study focus*

F_1 is not used here because we do not know the genotype of one of the flies. Only use F_1 and F_2 if both of the parents are known to be homozygous.

☑ *Study focus*

This test cross could also be done with females showing the dominant phenotype and males which are homozygous recessive. The results would be the same.

Summary questions

1 Explain the following terms: *monohybrid cross, dominant allele, recessive allele, homozygous, heterozygous* and *test cross.*

2 Plants of the species *Brassica rapa* normally grow a long stem, but some have a rosette habit in which the stem does not grow at all. When pure bred long-stemmed plants are crossed with pure bred rosette plants, all the offspring have long stems. When these plants are interbred three quarters of the offspring were long-stemmed and one quarter of them were rosette. Using the symbols **A** for long stem and **a** for rosette, draw a genetic diagram to explain these results.

3 Use a genetic diagram to explain how you would find out the genotype of a long-stemmed plant of the species *B. rapa*.

Learning outcomes

On completion of this section, you should be able to:

- define the terms *codominance* and *sex linkage*
- use genetic diagrams to solve problems involving codominance and sex linkage.

◯◯ Link

Now answer Summary question 2 to see the results of a test cross involving codominant alleles.

Summary questions

1 Explain what is meant by the term *codominance*.

2 A pink-flowered four o'clock plant was crossed with a white-flowered plant. Use a genetic diagram to show the outcomes of this test cross.

3 One of the blood group systems is the MN system. The two alleles, *GYPA*^{M} and *GYPA*^{N} are codominant. Draw a genetic diagram to show how two parents who both have the blood group MN may have children with different blood groups. State the probability that a child could inherit each of the blood groups.

Codominance

The four o'clock plant (Marvel of Peru), *Mirabilis jalapa*, has three different flower colours. If pure bred red-flowered plants are crossed with pure bred white-flowered plants then the F$_1$ plants have pink flowers. They are different from either of the two parental phenotypes. This looks like 'blending inheritance' that Mendel disproved with his experiments on peas. However, if these pink-flowered plants are interbred then the next generation gives all three colours in a 1:2:1 ratio. This is the same as the genotypic ratio in the monohybrid cross on page 94.

The reason for the intermediate colour (pink) is that both alleles contribute to the phenotype of the heterozygous plants. They are known as codominant alleles. The alleles are written as follows:

$$C^R = \text{red}$$
$$C^W = \text{white}$$

The letter **C** represents the gene; the alleles are represented by **C** with either the superscript ^{R} or ^{W}. This indicates that the alleles are neither dominant nor recessive.

The genetic diagram shows the inheritance of flower colour in the four o'clock plant through two generations.

	red-flowered plant	×	white-flowered plant
parental phenotypes			
parental genotypes	C^{R}C^R		C^{W}C^W
parental gametes	(C^R)	+	(C^W)
F$_1$ genotype		C^{R}C^W	
F$_1$ phenotype		all pink	
F$_1$ phenotypes	pink	×	pink
F$_1$ genotypes	C^{R}C^W		C^{R}C^W
F$_1$ gametes	(C^R), (C^W)	+	(C^R), (C^W)

		male gametes	
		(C^R)	(C^W)
female gametes	(C^R)	C^{R}C^R	C^{R}C^W
	(C^W)	C^{R}C^W	C^{W}C^W

F$_2$ genotypes and phenotypes	C^{R}C^R red	2C^{R}C^W pink	C^{W}C^W white

F$_2$ phenotypic ratio   1 red:2 pink:1 white

This shows that if alleles are codominant more variation appears as there are three flower colours not two, as would be the case if the allele for red was dominant to the allele for white.

Flower colour depends on the production of pigments in the petals. Enzymes catalyse the conversion of a colourless substance into the red pigment. The enzymes coded for by two **C**^{R} alleles are required to produce the red pigment. The **C**^{W} allele codes for an inactive enzyme. Where there is one **C**^{R} allele there is only enough enzyme molecules produced in the cells to make some of the red pigment which gives the petals a pink colour.

Sex linkage

Among the flies in Morgan's collection were some that had white eyes rather than the usual wild type red eyes. When Morgan investigated the inheritance of eye colour the results did not match the prediction that there would be a $3:1$ ratio in the F_2 as with the inheritance of wing length.

When pure-bred white-eyed female fruit flies were crossed with pure-bred red-eyed male fruit flies, the next generation were not red-eyed as might have been predicted. Instead all the females were red-eyed and all the males were white-eyed. There were *no* red-eyed males and *no* white-eyed females. The expression of eye colour had swapped sexes. When this F_1 generation were crossed the next generation had all combinations of sex and eye colour in approximately equal numbers.

The inheritance of this gene was similar to the inheritance of the X chromosome. The gene for eye colour is on the X chromosome. There is no equivalent gene on the Y chromosome. This means that females have two copies of this gene in the same way as with genes that have their loci on autosomes. Males only have one copy as they only have one X chromosome. To show inheritance of a sex-linked gene it is usual to include the X and Y chromosomes and put the symbols for the alleles as superscripts. The table shows all the different genotypes for the inheritance of eye colour in *Drosophila*.

Males are hemizygous. They have one allele of each of the genes on the X chromosome. These alleles are always expressed as there is no chromosome homologous to their X chromosome. Females which are heterozygous are described as carriers when the mutant allele is recessive as in this example.

Males		Females	
genotype	phenotype	genotype	phenotype
X^RY	red eyes	X^RX^R	red eyes
X^rY	white eyes	X^RX^r	red eyes
		X^rX^r	white eyes

Note the following features of sex linkage:

- the recessive phenotype is more common in males than in females
- males cannot inherit the recessive condition from their male parent, but females can
- females can be heterozygous and therefore carriers of the condition, males can never be carriers
- males inherit their Y chromosome from their father and this does not have the loci that are found on the X chromosome.

☑ *Study focus*

If a question says that males and females of a certain species were crossed, that does not necessarily mean that the gene involved is sex-linked. You need to look for other clues. See Summary question 6 below and Question 7 on page 109.

parental phenotypes   red-eyed male   ×   white-eyed female
parental genotypes X^RY X^rX^r
parental gametes $(X^R), (Y)$ + (X^r)

		female gametes
		(X^r)
male gametes	(X^R)	X^RX^r
	(Y)	X^rY

F_1 genotype(s) X^rY X^RX^r
F_1 phenotype(s)   white-eyed male   red-eyed female
F_1 genotypes X^rY × X^RX^r
F_1 gametes $(X^r), (Y)$ + $(X^R), (X^r)$

		female gametes	
		(X^R)	(X^r)
male gametes	(X^r)	X^RX^r	X^rX^r
	(Y)	X^RY	X^rY

F_2 genotypes and phenotypes X^RY X^rY X^RX^r X^rX^r
                         red-eyed   white-eyed   red-eyed   white-eyed
                          male   male   female   female

Summary questions

4 Explain the term *sex linkage*.

5 Explain how sex is determined in fruit flies and in humans.

6 Suggest why sex linkage is seen in fruit flies but not in plants, such as the four o'clock plant, *Mirabilis jalapa*.

7 a Find out about the inheritance of haemophilia and red-green colour blindness in humans.

   b Show, using a genetic diagram, how a man cannot inherit haemophilia from his father, but can pass it to his grandson.

So far we have looked at the inheritance of one gene with two alleles. Humans have about 20 000 genes that code for proteins, fruit flies have about 14 000 and the number in *B. rapa* is 43 000. Let's start with looking at the inheritance of two genes. You might predict that since fruit flies have four different chromosomes, that there are about 4000 genes on each and that the inheritance pattern will not be simple.

Dihybrid cross

Wild type fruit flies have long wings and grey bodies. There are some that have short wings and black bodies. Pure bred flies of both types were crossed and all the offspring had the wild type phenotype. When these F_1 flies were crossed all combinations of features appeared in the offspring, in the following numbers:

- wild type (long wings and grey body), 650
- long wings and black body, 198
- short wings and grey body, 225
- short wings and black body, 68.

This approximates to a ratio of $9:3:3:1$. But if you look you can see that the ratio of long wings to short wings is about $3:1$ and the ratio of grey body to black body is also about $3:1$. This suggests that the two genes have segregated independently of one another and are therefore on different chromosomes.

The genetic diagram shows what happens in this dihybrid cross which involves the inheritance of two genes.

parental phenotypes	wild type male ×	female with short wings and black body
parental genotypes	**WWGG**	**wwgg**
parental gametes	(WG) +	(wg)

F_1 *genotype(s)*	**WwGg**	
F_1 *phenotype(s)*	all long wings and black body	
F_1 *genotypes*	**WwGg** ×	**WwGg**
F_1 *gametes*	(WG) (Wg) (wG) (wg)	(WG) (Wg) (wG) (wg)

		female gametes			
		(WG)	(Wg)	(wG)	(wg)
male gametes	(WG)	WWGG	WWGg	WwGG	WwGg
	(Wg)	WWGg	WWgg	WwGg	Wwgg
	(wG)	WwGG	WwGg	wwGG	wwGg
	(wg)	WwGg	Wwgg	wwGg	wwgg

F₂ genotypes and phenotypes

genotypes	phenotypes	proportions	
WWGG	long wing, grey body	$\frac{1}{16}$	
WWGg		$\frac{2}{16}$	$\frac{9}{16}$
WwGG		$\frac{2}{16}$	
WwGg		$\frac{4}{16}$	
WWgg	long wing, black body	$\frac{1}{16}$	$\frac{3}{16}$
Wwgg		$\frac{2}{16}$	
wwGG	short wing, grey body	$\frac{1}{16}$	$\frac{3}{16}$
wwGg		$\frac{2}{16}$	
wwgg	short wing, black body	$\frac{1}{16}$	$\frac{1}{16}$

F₂ phenotypic ratio

9 long wing, grey body : 3 long wing, black body : 3 short wing, grey body : 1 short wing, black body

The Punnett square shows that there are nine different genotypes, but because of dominance there are only four different phenotypes in the ratio 9 : 3 : 3 : 1, which approximates to the numbers given opposite.

Mendel called this **independent assortment**, which means that the two genes segregate in meiosis independently of one another. This is because they are on different chromosomes. You can see this in Figure 3.5.1 and also you can use chromosome models to model this (see page 87).

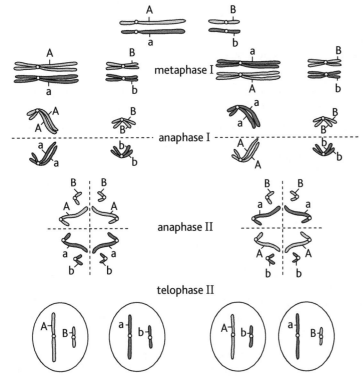

Figure 3.5.1 *Independent assortment of the alleles of two genes, **A/a** and **B/b***

✓ *Study focus*

Working out the genotypes in the F₂ should show you why using a Punnett square is so useful. Do not use criss-cross lines for this.

✓ *Study focus*

Figure 3.5.1 shows what happens as a result of two pairs of homologous chromosomes aligning on the equatorial plate during metaphase I. In some cells the maternal and paternal chromosomes will align as shown on the left and in others they will align as on the right. There is a 50% chance that paternal and maternal chromosomes will align in each of these ways.

Summary questions

1 Explain the meaning of the terms *dihybrid inheritance* and *independent assortment*.

2 *Brassica rapa* plants have either green leaves or yellow leaves. The allele for green leaves is dominant. Use genetic diagrams to show the outcomes in the F₁ and F₂ if a pure bred plant with long stems and green leaves is crossed with a rosette plant with yellow leaves.

3 The Punnett square shows that there are nine different genotypes in the F₂ of the dihybrid cross. Suggest how it would be possible to carry out a dihybrid cross so that each of those genotypes had a different phenotype.

The next genetic diagram shows what happens in a test cross on a fruit fly that is heterozygous for both gene loci.

Link

If you find independent assortment a difficult concept to understand, use the chromosome models recommended on page 87 and attach sticky labels to show the alleles of the two genes while referring to Figure 3.5.1.

parental phenotypes	wild type male	×	female with short wings and black body
parental genotypes *parental gametes*	**WwG** (WG), (Wg), (wG), (wg)		**wwgg** (wg)

		female gametes
		(wg)
male gametes	(WG)	**WwGg**
	(Wg)	**Wwgg**
	(wG)	**wwGg**
	(wg)	**wwgg**

offspring genotypes and phenotypes

genotypes	phenotypes	proportions
WwGg	long wing, grey body	$\frac{1}{4}$
Wwgg	long wing, black body	$\frac{1}{4}$
wwGg	short wing, grey body	$\frac{1}{4}$
wwgg	short wing, black body	$\frac{1}{4}$

offspring phenotypic ratio   1 long wing, grey body: 1 long wing, black body: 1 short wing, grey body, 1 short wing, black body

The ratio $1:1:1:1$ is the test cross ratio for a dihybrid cross with dominance at each gene locus.

Study focus

Remember that there is a 50% probability that two pairs of homologous chromosomes will align themselves on the metaphase plate with the dominant alleles of the two genes together.

Study focus

You must state a null hypothesis before you use the chi-squared test to analyse your data.

If you cannot do any practical genetics with living organisms then it is good to try some computer simulations or model monohybrid and dihybrid crosses to generate some data of your own.

There are free simulations available on the internet that will generate data for you. Alternatively, you can use this simple way to simulate monohybrid and dihybrid crosses.

Monohybrid F_1 cross. Take 200 beads, 100 of one colour and 100 of a different colour. These represent the alleles from 50 F_1 individuals that are heterozygous. (Anything suitable will do in place of beads. But they must all be the same size and shape so you cannot tell them apart by touch.) Decide which colour is dominant and predict the result that you will expect in the 'F_2'. Devise a table to record the results. Place all the beads in a bag and shake them up. Remove two beads and record the colours of the beads. Each pair of beads represents a genotype. Record the genotype and phenotype. Replace the beads in the bag, shake and pick another pair of beads. You must pick the beads at random, so do not use a transparent bag!

When you have 50 'F_2 offspring', total the numbers of each genotype and phenotype. Record the totals and carry out a chi-squared test (see page 104) to see if the difference between the observed results (phenotypes) and your predicted result is significant or not.

Dihybrid cross. Repeat the same procedure but use four different colours with 100 beads of each colour. Put two of the colours into one bag (gene 1) and put the other two colours in another bag (gene 2).

Summary questions

4 Describe why the dihybrid F_2 ratio is $9:3:3:1$ whereas the test cross ratio is $1:1:1:1$.

5 The results of these crosses are never as exactly predicted by the ratios. Suggest why this is the case.

6 Make a table similar to the one on page 73 to show the effects of genes in *D. melanogaster*. You can find chromosome maps of *Drosophila* on the internet.

To make the genotype of each offspring pick two beads from each bag. Record the genotype and phenotype and then return the beads to their respective bags and pick again. You could make this more interesting by having codominance at one of the gene loci.

Genetic ratios

You should be able to identify genetic ratios from the results that you are given in a question. This table should help you.

✓ *Study focus*

Read this table carefully. In each case go back to the relevant pages from this guide and make sure you understand the descriptions given in this table.

Ratio	Description of the cross	Page number
3:1	F_2 from a monohybrid cross with dominance; the parental generation was homozygous	94
1:2:1	F_2 from a monohybrid cross with codominance; the parental generation was homozygous	96
1:1	offspring from a monohybrid test cross with 'unknown' being heterozygous	95
9:3:3:1	F_2 from a dihybrid cross with dominance at both loci; the parental generation was homozygous for both gene loci	99
1:1:1:1	offspring from a dihybrid test cross with 'unknown' being heterozygous at both loci; dominance at both loci	100
1:1:1:1	F_2 from a cross involving sex linkage where the females of the parental generation were homozygous recessive and the males were hemizygous dominant	97

Summary questions

7 Usually, tomato plants have purple stems, although in some varieties the purple pigment is missing and the stem is green. Most tomato plants have 'cut' leaves, but some have so-called 'potato' leaves.

The table shows the numbers and phenotypes of offspring, resulting from four separate crosses of tomato plants.

✓ *Study focus*

Remember to write out the genetic diagrams for Questions 2 and 7 in full, following the instructions on page 93.

Parental phenotypes	Number and phenotypes of offspring			
	purple stem, cut leaves	purple stem, potato leaves	green stem, cut leaves	green stem, potato leaves
1 purple stem, cut leaves × green stem, potato leaves	354	0	367	0
2 purple stem, cut leaves × green stem, cut leaves	693	241	0	0
3 purple stem, potato leaves × green stem, cut leaves	60	71	56	67
4 purple stem, cut leaves × green stem, cut leaves	323	101	308	109

a Calculate the genetic ratio for each of these crosses.

b Identify the genes involved and explain the relationship between the alleles of each gene. Choose suitable symbols for the alleles.

c Use genetic diagrams to explain the results of each of the crosses.

Genotype	Phenotype
I^A I^A	blood group A
I^A I^O	
I^A I^B	blood group AB
I^B I^B	blood group B
I^B I^O	
I^O I^O	blood group O

☑ Study focus

The probability of these parents having a baby with any one of the blood groups is always 1 in 4 regardless of the blood groups of children that they have already.

Did you know?

Blood typing is essential so people get given blood of the same group during a transfusion. In emergencies, when there is shortage of blood, people with AB blood can receive blood of any other groups; blood of type O can be given safely to everyone, although there are other blood group systems (e.g. Rhesus) that must be tested as well.

Multiple alleles

So far we have looked at genes with two alleles. The pairs of alleles have either been dominant and recessive or have been codominant. Many genes have more than two alleles. There are three alleles at the human ABO blood group locus. The gene locus is on chromosome 9 of the human genome. The gene controls the production of antigens on the red blood cells. They are called antigens because if red cells from one person are injected into an experimental animal they are detected as 'foreign' and stimulate the production of antibodies. If red blood cells are injected into another human then sometimes they stimulate the same response, but not always. This is why blood must be typed before a blood transfusion is carried out.

The gene locus is known as I and it has three alleles, I^A, I^B and I^O. I^A and I^B are codominant and both are dominant to I^O. There are six genotypes and four phenotypes as shown in the table on the left.

Any individual can only have two of these alleles as humans are diploid with two copies of chromosome 9 in each cell.

The genetic diagram shows how a man who has blood group A and a woman who has blood group B can have children with all four blood groups.

	male blood group A	×	female blood group B
parental phenotypes			
parental genotypes	I^{A}I^O		I^{B}I^O
parental gametes	(I^A) , (I^O)	+	(I^B) , (I^O)

		male gametes	
		(I^A)	(I^O)
female gametes	(I^B)	I^{A}I^B	I^{B}I^O
	(I^O)	I^{A}I^O	I^{O}I^O

offspring genotypes	I^{A}I^O	I^{A}I^B	I^{B}I^O	I^{O}I^O
offspring phenotypes	blood group A	blood group AB	blood group B	blood group O

The phenotypic ratio is $1:1:1:1$. In human genetics it is more usual to give the probability of a child inheriting these blood groups. In this family there is a 0.25 (25% or 1 in 4) probability that a child will have any one of these blood groups.

Epistasis

Some features are controlled only by one gene as is the case with the ABO blood group. There are features that are controlled by more than one gene. Often these genes interact with each other. **Epistasis** is the interaction of two or more gene loci in the control of a phenotypic feature. The genes involved will show independent assortment if they are on different chromosomes. In such cases, however, the phenotypic ratios are different to those expected with independent assortment (see page 99).

Flower colour in blue-eyed Mary, *Collinsia parviflora*, is a good example, as the pigment which gives the petals their colour is produced by reactions catalysed by enzymes.

In *C. parviflora* there are two such reactions that occur in series as shown on the right.

Enzyme 1 is coded by the gene **A/a** and enzyme 2 is coded by the gene **B/b**. If the genotype of the plant has dominant alleles of both genes then the flowers are blue. In plants that are homozygous recessive, **aa**, it does not matter whether enzyme 2 is present or not as the substance Y will not be produced and the flowers are white. If allele **A** is present then Y will be produced. In plants that have allele **A** but recessive, **bb**, then Y will not be converted to Z and the plant will have magenta flowers. This cross does not give a 9:3:3:1 ratio as in the dihybrid cross on page 99.

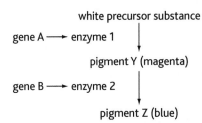

Figure 3.6.1 *These two enzymes catalyse reactions to produce a flower pigment*

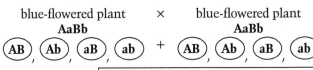

	female gametes			
	(AB)	(Ab),	(aB)	(ab)
(AB)	AABB	AABb	AaBB	AaBb
(Ab)	AABb	AAbb	AaBb	Aabb
(aB)	AaBB	AaBb	aaBB	aaBb
(ab)	AaBb	Aabb	aaBb	aabb

male gametes

offspring genotypes and phenotypes

genotypes	phenotypes	proportions	
AABB	blue	$\frac{1}{16}$	
AABb		$\frac{2}{16}$	$\frac{9}{16}$
AaBB		$\frac{2}{16}$	
AaBb		$\frac{4}{16}$	
AAbb	magenta	$\frac{1}{16}$	$\frac{3}{16}$
Aabb		$\frac{2}{16}$	
aaBB	white	$\frac{1}{16}$	
aaBb		$\frac{2}{16}$	$\frac{4}{16}$
aabb		$\frac{1}{16}$	

offspring phenotypic ratio
9 blue : 4 white : 3 magenta

Both **A** and **B** are needed to produce blue, but in **A-bb** magenta is produced as no functioning enzyme 2 is made. But when there are only recessive alleles (**aa**), neither **B** (to give blue) nor **b** (to give magenta) is expressed.

In this example the gene **A/a** is epistatic over gene **B/b** since if the genotype is **aa** (homozygous recessive) gene **B/b** cannot be expressed. Gene **B/b** is said to be the hypostatic locus. This is **recessive epistasis**. An epistatic gene may also act by inhibiting the action of another gene, perhaps by inhibiting its transcription or by coding for an inhibitor of an enzyme. This type is known as **dominant epistasis**. Epistasis reduces variation, since there is one less phenotype compared with the situation where each gene controls a different feature as you saw on page 99.

Summary questions

1 Define the terms *multiple alleles*, *epistasis*, *recessive epistasis* and *dominant epistasis*.

2 A couple are blood group B and blood group A. Use genetic diagrams to predict the blood groups of their children.

3 Suggest how genes for cell surface antigens on red blood cells are expressed in cells that have no nuclei.

4 Draw a genetic diagram to predict the results of crossing a magenta-flowered plant of *C. parviflora* with a plant homozygous recessive for both gene loci.

5 What effects do dominant epistasis and recessive epistasis have on phenotypic variation? Explain your answer.

Learning outcomes

On completion of this section, you should be able to:

■ Analyse the results of a genetic cross by:

   ■ stating the null hypothesis

   ■ completing a chi-squared test in a table

   ■ using a probability table to assess the significance of the results.

 Link

The type of data collected in genetic studies is called categoric data. This is because you count the number of individuals in different categories, such as green, cut and purple potato. Read more about this on page 124.

☑ *Study focus*

You can find programs to carry out the chi-squared test for you, but as you may be asked to do a chi-squared test you should learn how to set out the calculation in this table.

If you carry out breeding experiments or run a computer simulation on inheritance, you will not gain results that exactly fit any of the ratios we have obtained so far. Does this matter? In 1900, Karl Pearson (1857–1936) developed the chi-squared test as a 'goodness of fit' test to check the significance of differences between observed and expected results when using categoric data.

Tomato plants have indented leaves, known as 'cut'. Some tomato plants have leaves shaped like those of potatoes. Pure bred tomato plants with cut leaves and purple stems were crossed with pure bred plants with potato leaves and green stems. All the F_1 generation had cut leaves and purple stems. These F_1 plants were test crossed against tomato plants showing the recessive phenotype – potato leaves and green stems. The test cross offspring showed the following numbers of plants in each of four phenotypes:

purple, cut	purple, potato	green, cut	green, potato
70	91	86	77

The ratio of phenotypes *expected* in the test cross offspring of a dihybrid cross such as this is $1:1:1:1$ assuming that the two genes are on different chromosomes and independent assortment has happened during meiosis I to form the gametes of the F_1 plants.

The **null hypothesis** is the hypothesis that states there is no significant difference between the observed and expected results.

The chi-squared test (χ^2) is used to find out if these results are significantly different from the expected results.

The formula for calculating χ^2 is:

$$\chi^2 = \sum \frac{(O-E)^2}{E}$$

Σ = sum of....
O = observed value
E = expected value

The figures are put into this table.

Categories	O	E	O–E	(O–E)²	(O–E)²/E
purple cut	70	81	−11	121	1.49
purple potato	91	81	10	100	1.23
green cut	86	81	5	25	0.31
green potato	77	81	−4	16	0.20
totals	324	324		$\chi^2 =$	3.23

The statistic, χ^2, has the value of 3.23 in this example. To find out if this is significant or not, we need to look at a table of probabilities. This table shows how to do this.

Degrees of freedom	Distribution of χ^2							
	← increasing values of p decreasing values of p → probability, p							
	0.99	0.90	0.50	0.10	0.05	0.02	0.01	0.001
1	0.00016	0.016	0.46	2.71	3.84	5.41	6.64	10.83
2	0.02	0.21	1.39	4.61	5.99	7.82	9.21	13.82
3	0.12	0.58	2.37	6.25	7.82	9.84	11.35	16.27
4	0.30	1.06	3.36	7.78	9.49	11.67	13.28	18.47

$p > 0.90$	$p > 0.05$	$p < 0.05$	$p < 0.01$	$p < 0.001$
result is 'dodgy' = too good!	result is not significantly different from expected outcome	result is significantly different from expected outcome	highly significant	very highly significant

Note: > = greater than; < = less than.

To use the table we need to know how many degrees of freedom (df) there are. In this example, you score the plants using a tally chart (see page 126). If a plant has cut and potato leaves there are three other categories *that it could be scored as*. This represents the degrees of freedom which in this case is 3.

The next step is to read across the table at 3 df until you come to the column for $p = 0.05$. This is the probability that we will get this result 5% of the time or once in every 20 times we carry out the investigation. The column and row intersect at 7.82 which is the **critical value**. Our result, 3.23, is *less* than this critical value, which means that the result is not significantly different from the expected outcome and we can accept the null hypothesis. The decision to use a probability of 0.05 or 5% was arbitrary, but is now the accepted value in biological investigations.

It is more precise to say that the value of p is greater than 0.1 and less than 0.5. The probability of getting this result is therefore between 10% and 50% which means that the result is due to chance effects such as random fertilisation. The difference is not *statistically significant* and so the null hypothesis can be accepted. If the value for χ^2 is greater than the critical value then the probability of the results is *less than 0.05* and there is a *significant difference* between observed and expected. If so, the prediction is rejected, refined or the experimental procedure is reviewed to see if there are any errors.

Curled wings and spineless bristles are autosomal recessive features in *Drosophila*. Pure-breeding wild type flies with straight wings and normal bristles were crossed with pure-breeding flies with curled wings and spineless bristles. The F_1 all had straight wings and normal bristles. Female flies from the F_1 were test-crossed with males homozygous for both gene loci. The results were as follows overleaf.

☑ *Study focus*

In the exam you would only be given the top half of this table as the bottom half tells you how to interpret the values for χ^2.

☑ *Study focus*

The table of probabilities you will be given in the examination will be similar to the one on page 107, not the one on this page. To use the table, calculate the df by subtracting one from the number of categories in the table and remember to find the critical value at $p = 0.05$

🔗 *Link*

Autosomal means that the gene locus is on an autosome and not on a sex chromosome (see page 92).

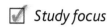

Study focus

Draw a table like that on page 104 to show how to derive the value for χ^2 of 275.76.

■ straight wings, normal bristles = 186; straight wings, spineless bristles = 18;

■ curled wings, normal bristles = 16; curled wings, spineless bristles = 180.

The ratio of phenotypes expected in a cross such as this is $1:1:1:1$.

The null hypothesis is that there is no significant difference between the observed and expected results.

The value for χ^2 is calculated as 275.76.

It is obvious that the chi-squared value is going to be large because the observed numbers are so different from the expected numbers. Putting the χ^2 value of 275.76 into the table at df = 3, we see that the probability of these results departing significantly by chance from the expected ratio is much less than 0.001 (0.1%). This is such a low probability that we can reject the null hypothesis and conclude that these two genes do *not* assort independently and cannot be on different chromosomes. The likely explanation is that they are on the same chromosome, in fact chromosome 3.

If you are asked to do a chi-squared test, these are the steps you follow.

1 Analyse the information to see what type of inheritance pattern there is: is it monohybrid, dihybrid, sex-linked, etc.
2 Determine the expected ratio if that has not been done for you.
3 Write a null hypothesis. This always starts 'there is no significant difference between …'
4 Draw a chi-squared table as shown on page 104. The number of columns is always six, but the number of rows depends on the number of classes of data. For a monohybrid cross with two phenotypes among the offspring you need three rows; for a dihybrid cross with four phenotypes among the offspring you need five rows in your table. (One row is needed for the headings of the columns.)

 Link

Try doing chi-squared tests on your own data from the simulation on pages 100–101.

5 Complete the table by:
  a calculating the expected results using the predicted ratio
  b calculating the difference between observed and expected results
  c squaring the difference (to remove signs)
  d dividing the square of the difference by the expected numbers for each class (this takes into consideration the size of the sample)
  e adding up the results of the final column to give the statistic, χ^2
  f deciding on the degrees of freedom (number of classes − 1)
  g finding the 0.05 probability (p) in the table of probabilities for the appropriate df
  h identifying the critical value
  i checking to see if the χ^2 value is greater or less than the critical value.

If it is less then there is no significant difference between the predicted and expected results and the null hypothesis is accepted.

If it is greater then there is a significant difference between the predicted and expected results and the null hypothesis is rejected. The result can be expected less than once in 20. Express your conclusions in terms of the probability of getting the result by chance, e.g. $p < 0.1$ but > 0.05.

Summary questions

1 The results of certain types of experiments can be analysed with the χ^2 test – what do these experiments have in common?

2 What type of data can be analysed using the χ^2 test?

3 Given the formula:

$$\chi^2 = \sum \frac{(O-E)^2}{E}$$

what are the row and column headings in the table for calculating χ^2?

4 How do you calculate degrees of freedom?

5 What does the χ^2 value mean?

6 What is the importance of the 5% level (or 0.05)?

7 What does p stand for?

8 What do you conclude if the value for p is greater than (>) 0.05 (5%)?

9 What do you conclude if the value of p is less than (<) 0.05 (5%)?

10 What do you conclude if the value of p is greater than (>) 0.9 (90%)?

In questions 11 to 13 you should use this table of probabilities for the chi-squared test.

Degrees of freedom	distribution of χ^2						
	probability, p						
	0.90	0.50	0.10	0.05	0.02	0.01	0.001
1	0.02	0.46	2.71	3.84	5.41	6.64	10.83
2	0.21	1.39	4.61	5.99	7.82	9.21	13.82
3	0.58	2.37	6.25	7.82	9.84	11.35	16.27
4	1.06	3.36	7.78	9.49	11.67	13.28	18.47

✓ *Study focus*

Make sure you follow the instructions in this section when using the table. The more practice you have the better!

11 A student investigated inheritance in fruit flies. The student crossed some vestigial winged fruit flies with some wild type (long winged) fruit flies. All the F_1 fruit flies were long winged. The F_1 flies were interbreed with these results in the F_2 generation:

long winged   164   vestigial winged   36

a State the phenotypic ratio that the student should expect for the F_2 results.

b Give the null hypothesis for this investigation.

c Use the chi-squared test and the table of probabilities to find if the results of this cross differ significantly or not from the expected results.

d What conclusion can you make from your answer to **c**?

12 Plants of Marvel of Peru have either red, white or pink flowers. Pink-flowered plants were bred amongst themselves to give the following results:

red-flowered plants   69
white-flowered plants   193
pink-flowered plants   80

a Use your knowledge of genetics to give a null hypothesis for this investigation; use the chi-squared test to analyse these results. Show your working.

b State with a reason whether or not your analysis supports the null hypothesis you made.

c What conclusion can you make from your answer to **b**?

13 Another student carried out a breeding experiment to investigate the inheritance of wing length in fruit flies. The F_2 results were:

long winged   65   vestigial winged   21

What conclusions could you make concerning these results? Explain your answer in full.

 Link

In answering Question 12, look at Section 3.4

Learning outcomes

On completion of this section, you should be able to:

- make predictions about the outcomes of genetic crosses

- analyse pedigrees to determine genotypes and make predictions

- describe patterns of inheritance and explain them using genetic diagrams.

✓ Study focus

Make sure you set out your genetic diagram *in full*. Use the information on page 93 to help you.

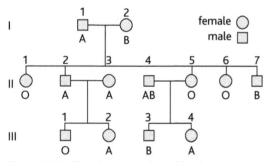

Figure 3.8.1 *The inheritance of ABO blood groups in a family*

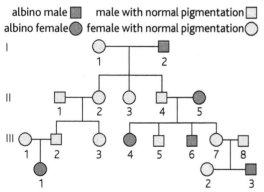

Figure 3.8.2 *The inheritance of albinism*

Pedigree diagrams

Questions on genetics are sometimes set in the context of a pedigree diagram or family tree. Geneticists use these when investigating a characteristic as it often gives information about the inheritance pattern. Nancy Wexler (1945–) discovered the position of the gene responsible for Huntington's disease, a serious neurological disorder, by studying all the descendants of Mary Soto who died from the disease in the early 1800s. Many of these descendants live in fishing villages around Lake Maracaibo in Venezuela.

The best way to be sure that you have learnt the genetics in this chapter is to interpret some data and make genetic diagrams. In the pedigree diagrams of the genetic conditions that follow the following symbols are used:

- circle = female; square = male

- shaded circle or square = person with symptoms of a genetic condition; unshaded circle or square = person who does not show any symptoms of the condition.

Individuals are identified by generation and number, e.g. I–1 and III–5.

The ABO blood groups

1 Study Figure 3.8.1 carefully and work out the genotypes of all the people in the pedigree diagram. You can start this by identifying those who are AB and O. Their genotypes must be $I^A I^B$ and $I^O I^O$. Now look at the parents of people who are $I^O I^O$ and you know that they must have the recessive allele, I^O, and be heterozygous if they are blood group A or B. With this information you should be able to work out the genotypes of all the people in the pedigree.

2 Draw genetic diagrams to show the blood groups of any children that III–3 and III–4 might have and predict the probabilities of each.

Albinism

This pedigree is different to the one above because there are just two phenotypes and most people do not show the condition. This suggests that the condition is recessive and if you look carefully you can see that both males and females have inherited the condition from parents who do not show the condition. This suggests that it is an **autosomal recessive** condition.

3 Work out the genotypes of all the people in the pedigree, using **A** as the symbol for the dominant allele and **a** as the symbol for the recessive allele. You may not know whether some of the people who do not show albinism are homozygous dominant or heterozygous. If this is the case use the dash for the allele you do not know (see the test cross on page 95).

4 Known carriers of genetic conditions are indicated by dots in the centres of the unshaded circles or squares. Explain which people in the pedigree for albinism carry the recessive allele. A number of people may carry the recessive allele but we cannot tell from the pedigree alone who they are. Identify those who may be carriers.

Huntington's disorder

This condition is inherited in a very different way to albinism. Notice that both males and females in the pedigree have inherited the condition from a parent who had it. In the family used for this pedigree analysis the disorder does not skip a generation. This suggests that it is an **autosomal dominant** condition; the allele for Huntington's, **H**, is the dominant allele and **h** is the normal allele.

One complication with a pedigree analysis of Huntington's is that it develops late in life as a degenerative disease. Often people pass on the condition before they know they have it and some people inherit the disorder from a parent who died young without having developed the condition.

5 Work out the genotypes of all the people in the pedigree. If you are unsure possibly because they are too young, indicate this by using question marks for unknown alleles.

6 State the probability that III–1 and III–5 have the dominant allele for Huntington's disorder.

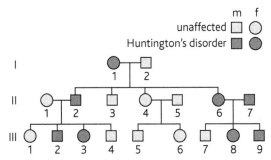

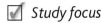

Figure 3.8.3 The inheritance of Huntington's disorder

> **Did you know?**
>
> Nancy Wexler's work on pedigree analysis located the gene for Huntington's disorder on chromosome 4. Search for 'Wexler Huntington's' to learn more.

> ☑ **Study focus**
>
> The genotype HH is never found. All people with Huntington's are heterozygous.

Haemophilia

Haemophilia is an inherited disorder in which the time taken for blood to clot is very long. All those with the disorder in the pedigree diagram are males and none of them have inherited the disorder from their fathers. This suggests it is a sex-linked recessive condition.

7 Draw a genetic diagram to show how person III–6 inherited haemophilia. Use **H** for the normal allele and **h** for the allele for haemophilia. Draw a genetic diagram to show the probability that III–4 and III–3 will have another child with haemophilia. What is the evidence from the pedigree diagram that haemophilia is a recessive condition?

8 What are the probabilities that a man will transmit haemophilia to **i** his son; **ii** his daughter; **iii** his grandson?

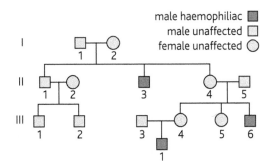

Figure 3.8.4 The inheritance of haemophilia

> ☑ **Study focus**
>
> Remember to include the X and Y chromosomes in your answers to Question 7; use superscripts for the alleles.

> **Did you know?**
>
> Geneticists contribute to many research projects, such as the Barbados National Cancer Study, investigating family links in the incidence of breast cancer.

Most of these questions give you information about inheritance patterns in prose. You have to interpret the information in order to draw genetic diagrams to provide an answer. In the examination, you will be given the symbols to use for the alleles so make sure you use them and do not invent your own.

1   a   Define the terms *gene* and *allele*.
    b   i   When pure bred tomato plants with cut leaves are crossed with pure bred tomato plants with potato leaves, all the offspring have cut leaves. Using **D** for the allele for cut leaves and **d** for the allele for potato leaves, draw a genetic diagram to explain why all the F_1 plants have cut leaves.
        ii Use a genetic diagram to predict the phenotypic ratio in the F_2 if F_1 plants are bred amongst themselves.
    c   A tomato breeder has a plant with cut leaves, but is not sure whether it is homozygous dominant or whether it is heterozygous.
        Show, by using genetic diagrams, how you could determine the genotype of a plant with cut leaves.

2   In mice there is a gene that controls fur colour. There are four alleles of this gene:

C = full brown colour (wild type); c^{ch} = chinchilla; c^d = extreme dilution; **c** = albino (white).

The gene is not sex-linked. The alleles show the following relationships to each other:

    C is dominant to the other three alleles
    c^{ch} is dominant to c^d and **c**, but is recessive to **C**
    c^d is dominant to **c**, but is recessive to c^{ch} and **C**
    c is recessive to the other three alleles.

    a   Draw a genetic diagram to show what types of mice would be expected in the F_1 and F_2 generations when a pure bred wild type mouse is crossed with a pure bred chinchilla mouse.
    b   Draw a genetic diagram to show what you would expect if you crossed a wild type mouse heterozygous for extreme dilution with a chinchilla mouse heterozygous for albinism.
    c   Is it possible for a pair of mice to have offspring that show all four phenotypes: wild type, chinchilla, extreme dilution and albino? Draw genetic diagrams to illustrate your answer.
    d   With reference to your answers to parts **a** to **c**, explain the term *multiple alleles*.

3   Tomato plants have genes that determine total height of the plants and the colour of the fruit. The allele for tall height, **T**, is dominant over that for dwarfness, **t**. The allele for red fruit, **R**, is dominant over that for yellow fruit, **r**. The gene loci for these two features are on different chromosomes.
    a   A test cross was made between two tomato plants.
        i   The possible genotypes of the gametes of the plant chosen as the male parent were **RT**, **Rt**, **rT** and **rt**. State the genotype of this plant.
        ii State the genotype of the plant chosen as the female parent. Explain your answer.
        iii State the phenotypes that you would expect among the test cross offspring. Give the ratio that you would expect.
    b   A cross was made between the same plant that was used as the male parent in the test cross and another with the genotype **rrTt**. Draw a genetic diagram to show the genotypes and phenotypes of the offspring and the expected phenotypic ratio.

4   In cats, short hair, **H**, is dominant over long hair, **h**. The gene for hair length is not sex-linked. An allele, B^y, of another gene that is sex-linked, produces yellow coat colour. The allele B^b produces black coat colour. The heterozygous condition B^yB^b is tortoiseshell.
    a   Use genetic diagrams in your answers to the following questions.
        i   If a long-haired, black male cat mated with a yellow female cat homozygous for short hair, what kinds of kittens will be produced in the next generation?
        ii If the offspring were allowed to interbreed freely, what are the chances of obtaining a long-haired, yellow male?
    b   i   Use the inheritance of these two genes in cats to explain the terms *sex linkage* and *codominance*.
        ii Explain why it is not possible to have a tortoiseshell male cat.

5   In shorthorn cattle there is a gene that controls coat colour. The allele C^R gives a red colour, the allele C^W gives white. Cattle that are heterozygous for this gene have a coat that is described as roan – a light red colour. Cattle with horns are homozygous for

the allele, **p**. Cattle with the dominant allele, **P**, are hornless. Neither of these gene loci is sex-linked. The two gene loci are on different chromosomes.

a i A white cow with horns is mated with a bull that has a red coat and is homozygous for the hornless condition. Draw a genetic diagram to predict what you would expect in the F_1 generation.

   ii Cattle in the F_1 generation were mated among themselves. Draw a genetic diagram to predict what you would expect in the F_2 generation.

b Explain, using this example of shorthorn cattle, the effect of dominance and codominance on phenotypic variation.

6 Like cassava, clover plants are cyanogenic. This means that they release hydrogen cyanide from leaves and roots when they are damaged. A chemical known as linamarin is broken down releasing hydrogen cyanide. This happens very slowly at normal temperatures, but the enzyme linamarase in cell walls speeds up the reaction.

The substrate and the enzyme are formed as a result of two gene loci, gene **G/g** and gene **E/e**, which are found on different chromosomes. The allele, **G**, codes for the enzyme glucosyltransferase that catalyses the production of linamarin from a precursor substance; **g** codes for an inactive enzyme. The allele **E** codes for the production of linamarase; **e** codes for an inactive enzyme. The relationship between genes, enzymes, substrates and products is summarised below:

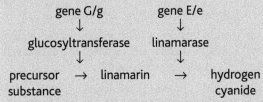

A pure breeding cyanogenic clover plant was crossed with a pure breeding non-cyanogenic plant. All the F_1 offspring were cyanogenic. When an F_2 generation was produced, three phenotypes were found:

- cyanogenic
- very slightly cyanogenic
- non-cyanogenic

in the ratio of 9:3:4.

a Draw a genetic diagram to show the genotypes of the parents, and the genotypes and phenotypes of the F_1 and F_2 generations of this cross.

b Explain why this cross does not show a typical dihybrid F_2 ratio of 9:3:3:1.

c As yet non-cyanogenic cassava plants have not been produced by selective breeding. Suggest why this is so.

7 A student carried out a genetic investigation with fruit flies, *Drosophila melanogaster*. Two characteristics were observed, body colour and wing shape. The dominant features were grey body and normal wings. The student carried out a test cross on fruit flies that were heterozygous for both gene loci.

The results were as follows:

grey body and normal wing, 83
black body and normal wing, 88
grey body and bent wing, 78
black body and bent wing, 74

The student concluded that the results showed that independent assortment had taken place.

a Carry out a chi-squared test on the experimental data to test the student's conclusion (use the table of probabilities on page 107).

b Show, by use of diagrams, how the behaviour of chromosomes during meiosis explains the independent assortment of the alleles for body colour and wing shape.
Use the symbols **G** and **g** for the alleles for body colour and **N** and **n** for the alleles for wing shape.

8 A student investigated the inheritance of the feature known as 'cut' wings in the fruit fly, *Drosophila melanogaster*. 'Cut' wing is controlled by a single gene. The student carried out two crosses, **A** and **B**, as shown below.

	Cross A		Cross B
parents	females with normal wings x males with 'cut' wings		males with normal wings x females with 'cut' wings
F_1	all normal wings		all males with 'cut' wings all females with normal wings
F_2 females	789 normal wings		356 normal wings 339 'cut' wings
F_2 males	391 normal wings 376 'cut' wings		342 normal wings 333 'cut' wings

a Draw a genetic diagram to explain the results of cross **A**. Use the symbol **N** for the allele for normal wings and n for the allele for 'cut' wings.

b Use your answer to a to explain how these results show that the allele for 'cut' wing is: **i** recessive; **ii** is sex-linked rather than being carried on an autosome.

c Give the genotypes of the parents, the F_1 and the F_2 in cross **B**.

2 Genetics, variation and natural selection

4.1 Principles of genetic engineering

Learning outcomes

On completion of this section, you should be able to:

- define the terms *genetic engineering* (*recombinant DNA technology*), *restriction enzyme* and *vector*

- explain how restriction enzymes and ligases are used in genetic engineering

- state the roles of vectors in genetic engineering.

Did you know?

DNA can also be 'shot' directly into cells rather than using a vector.

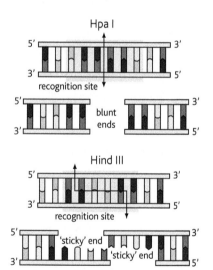

Figure 4.1.1 *The restriction sites in DNA for two restriction enzymes*

Did you know?

These enzymes are also known as restriction endonucleases because they cut within DNA molecules (*endo* – means within). They are used by bacteria as a defence against attack by viruses.

Moving genes

Genetic engineering is the process of genetic modification by means that are not possible using traditional breeding and artificial selection. It involves the removal of a gene, or genes, from one organism and placing the gene or genes into another organism. At one extreme this involves removing a gene from one species and transferring it to another species. But it can also involve taking a gene from an individual of a species and transferring it into other individuals of the same species. The DNA corresponding to a gene is obtained in one of several different ways. The DNA is then inserted into a **vector**, such as a virus, plasmid or liposome, which transfers the gene into host cells.

DNA is universal so it is possible to make transfers between widely different species – for example from a human into a bacterium, a jellyfish into a bacterium or a bacterium into a plant. All cells can 'read' the genetic code so the protein that is encoded by the gene can be produced in any cell.

These are some of the reasons for wanting to do this:

- bacteria and eukaryotic cells make special chemicals that are only produced in small quantities by other methods, e.g. enzymes for the food industry
- making crop plants resistant to diseases, pests and herbicides
- improving nutritional qualities of crop plants
- using animals to make human proteins for medicines that are difficult to obtain by other methods
- making bacteria absorb and metabolise toxic pollutants.

Restriction enzymes are used to cut genes from lengths of DNA. They cut across both strands of DNA at specific sites known as **restriction sites** as they are restricted to cut only at these sites.

Each restriction site has a specific nucleotide sequence which is **palindromic** as it reads the same in the 5' to 3' direction as in the 3' to 5' direction. Restriction enzymes are named after the bacteria from which they were first isolated.

Some restriction enzymes, such as *EcoR1*, cut DNA to give it free, unpaired 'ends'. These ends are known as 'sticky ends' because they will form base pairs with complementary sequences of bases. This is how a gene cut by a restriction enzyme can be inserted into a plasmid or into viral DNA. Others, such as *HPaI* and *HaeIII*, cut straight across DNA to give blunt ends.

Large quantities of DNA are produced by inserting the gene into bacteria and using the bacteria to replicate the gene. This is done by inserting the gene into a vector, such as a plasmid or a virus, which then takes it into the bacterium. The plasmid or viral DNA must have the same restriction site as the gene so that the same restriction enzyme can cut it to give a short length of complementary bases. DNA cut by a restriction enzyme, such as *HaeIII*, can have a length of nucleotides added to give it sticky ends. Copies of the gene are mixed with the plasmid. Some plasmids will take up the

112

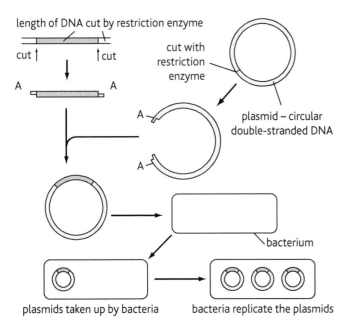

length of DNA cut by restriction enzyme

cut ↑   ↑ cut

A   A

cut with restriction enzyme

plasmid – circular double-stranded DNA

bacterium

plasmids taken up by bacteria   bacteria replicate the plasmids

Figure 4.1.2 Plasmids are used as vectors to insert genes into bacteria

gene by complementary base pairing of sticky ends; other plasmids will reform without taking up the gene. Hydrogen bonding between the 'sticky ends' attaches the two; the enzyme ligase is added to form the covalent phosphodiester bonds of the sugar–phosphate backbone of DNA.

Once the gene has become inserted into the plasmid **recombinant DNA** (rDNA) is produced. rDNA is DNA formed by combining DNA from two different sources.

The host bacteria are treated with calcium ions, then cooled and given a heat shock to increase the chances of plasmids passing through the cell surface membrane. The bacteria are now described as transformed as they contain foreign DNA.

Some plasmids do not take up the foreign gene and some bacteria do not take up plasmids. There are a variety of ways to identify the transformed bacteria from those that do not have the recombinant plasmids:

- plasmids contain antibiotic resistance genes; foreign genes are inserted into these resistance genes so making transformed bacteria sensitive to antibiotics
- plasmids contain fluorescence genes (from jellyfish); under UV light the transformed bacteria will fluoresce.

DNA polymerase in bacteria copies the plasmids; the bacteria then divide by binary fission so that each daughter cell has several copies of the plasmid. The bacteria transcribe and may translate the foreign gene. The bacteria are described as **transgenic**.

Bacteriophages are viruses that infect bacteria. They are used as vectors to deliver foreign genes as they attach to the cell wall of a bacterium and inject their DNA into the host cell.

Genetically-engineered bacteria are grown in large quantities to make specific proteins, such as enzymes for the food industry. However, this is not always possible as bacteria do not carry out the complex post-translation processes of eukaryotic cells to cut, fold and modify proteins by adding sugars. As a result genetic engineers use yeasts, plant cells or animal cells as the host cells for production of proteins.

∞ *Link*

Look back to the diagram of DNA on pages 63 and 64 to see the hydrogen bonds between the base pairs and the position of the phosphodiester bonds in each polynucleotide.

☑ *Study focus*

Plasmids are small rings of double-stranded DNA found in prokaryotes and in some eukaryotes.

Did you know?

Gene libraries can be made by cloning foreign genes within bacteria. This provides many copies of each gene for research purposes.

Summary questions

1 Define the terms *genetic engineering, restriction enzyme, ligase, vector, transformation* and *transgenic organism*.

2 Explain what is meant by the phrase 'DNA is universal'.

3 Describe in the form of bullet points how a gene may be removed from a eukaryotic cell and inserted into a bacterium.

4 Identify the advantages of genetic engineering.

5 Use the genetic dictionaries on page 67 to write a sequence of bases that will code for a decapeptide with restriction sites at either end. (A decapeptide has 10 amino acids.)

Learning outcomes

On completion of this section, you should be able to:

- explain the term *gene therapy*
- describe how gene therapy may be used in treatment for genetic disorders
- discuss the possible benefits and hazards of gene therapy.

∞ Link

Purines are component molecules of nucleic acids, refer to page 62 to remind yourself about them.

Did you know?

CF is the most common serious genetic disorder in people of Northern European origin affecting about 1 in 2000 live born infants.

Gene therapy is an application of the principle of genetic engineering. It involves the transfer of a normal functioning gene into a person who has a genetic disorder. There are a variety of methods of delivery of the genes into cells using:

- viruses that are taken up by specific cells
- liposomes that are small phospholipid bound vesicles
- plasmids that can be injected directly into cells.

SCID

Gene therapy was first used successfully to treat children with severe combined immunodeficiency syndrome (SCID). There are two forms of this condition. In the first the enzyme adenosine deaminase does not function. This happens in children who are homozygous recessive for the faulty allele of this gene. The enzyme is involved with the breakdown of purines, which are toxic to white blood cells. Children with the condition were treated by removal of the white blood cells known as T cells. These cells were given the dominant allele for this enzyme and then returned into the blood. This was first done successfully in 1990 and has been repeated since, but using modified stem cells from bone marrow rather than differentiated white blood cells.

The other form of the condition is X-linked SCID. A gene on the X chromosome codes for a cell membrane receptor for one of the immune system's cell signalling molecules. The gene therapy for this form of SCID was very similar and the dominant allele was inserted into stem cells.

Cystic fibrosis

Cystic fibrosis (CF) results from a mutation in the gene for a carrier protein (CFTR) that pumps chloride ions out of cells in the respiratory tract and in the alimentary canal. In people who are homozygous recessive for the disorder the mucus they produce is very thick and sticky and does not move easily. Research for many years has used viruses and liposomes as vectors to deliver the dominant allele into the cells of the respiratory tract using an inhaler. This has had mixed success as the vectors tend to stimulate inflammation and some clinical trials have had to be stopped.

Benefits and hazards

There are distinct benefits associated with gene therapy for SCID:

- it provides a treatment so that children no longer need to live in a sterile environment
- children no longer need to be treated with injections of adenosine deaminase and antibodies
- children can lead a normal life and, as far as anyone can predict, have a normal life span.

There are numerous problems with gene therapies:

- some boys with X-linked SCID have developed leukaemia; this happened because there is no control over where in the patient's genome the gene inserts; in this case the allele inserted into a gene controlling cell division

- it is difficult to get the allele delivered to all the cells in a tissue
- if an allele is taken up by a cell it may not be expressed
- only recessive conditions, such as CF, can be treated; it is not possible to switch off a dominant allele such as the one that is the cause of Huntington's disorder
- cells in the respiratory tract are short-lived and therefore the therapy for CF would have to be taken frequently
- the genes and the vectors prompt immune responses, which make the continued use of the therapy very difficult
- one person taking part in a trial in 1999 died and others have fallen ill as a response to the virus used as the vector
- many children with cystic fibrosis are born into families with no history of the disease although it may be that being autosomal recessive it has skipped several generations and affected family members who died young before CF was identified in the 1930s.

Other solutions

The type of gene therapy described above is **somatic gene therapy**. The solution to the problem of delivering genes to all the cells that require them could be overcome by placing the gene concerned into an egg or into a zygote so it is passed to every cell in the body. This is called **germ line gene therapy**.

You may be asked to discuss the practical, ethical and moral issues surrounding gene therapy.

- Practical issues include precautions taken during trialling, biological problems and solutions. Some people believe that gene therapy is at first sight a simplistic solution that involves many complex issues, such as the risks involved in inserting a gene without knowing exactly where it is going, leading to other health problems such as leukaemia. They also think it is very expensive and deflects research efforts from simpler and cheaper solutions, such as genetic screening.
- Ethical issues are of concern to the medical staff involved with the research and clinical trials. Before research can begin it must be approved by an ethics committee who will consider the risks and benefits involved in the research.
- Moral issues concern the wider society. People argue that we have no right to interfere with the genes that we have inherited; others reply that we have a moral imperative to apply the knowledge that we have about human genetics and apply any successful techniques to improve and extend the life of people with these life threatening genetic disorders.

Summary questions

1 Define the term *gene therapy*.

2 Find out about the current research into gene therapy for cystic fibrosis. Present your findings in a suitable form, such as a leaflet or e-mail campaign to raise awareness.

3 Explain the benefits and hazards of gene therapy.

4 Research different genetic disorders to find out how many may be treated in the future by gene therapy. Imagine you have to present your findings to a group, think about the main points you will highlight.

 Study focus

It is highly unlikely that gene therapy will offer any solutions to the most common disorders in humans, such as heart disease, high blood pressure, diabetes and Alzheimer's disease. This is because they are caused by the combined effects of variations in many genes, and, thus, injecting a single gene or even a few genes is unlikely to do any good.

Did you know?

Germ line gene therapy is not legal in any country. There are risks involved in a gene being passed on from one generation to the next.

 Study focus

Genetic screening involves carrying out DNA tests to find carriers of genetic disorders. There are tests for many of these: search for information about thalassaemia, Tay-Sachs and sickle cell anaemia to find out more about the benefits of genetic screening in reducing the number of cases of these very serious disorders.

 Study focus

Search for 'learn genetics' online to read more about the ethics of gene therapy.

Learning outcomes

On completion of this section, you should be able to:

- describe how genetic engineering is used to produce insulin

- outline the advantages and disadvantages of using insulin from genetically engineered cells

- explain why there is a increasing demand for insulin worldwide.

The concentration of your blood sugar is controlled by a variety of hormones. Of these, insulin is the only hormone that stimulates a decrease in the concentration of glucose in the blood. There are several other hormones that cause it to increase. Some people lose the ability to make insulin and as a result develop diabetes type 1.

Diabetes

Diabetes mellitus is a disorder caused by an inability to produce insulin or the inability of cells to respond to insulin. There are two types:

- diabetes type 1 – insulin-dependent
- diabetes type 2 – non insulin-dependent.

Type 1 diabetes is caused by an inability to secrete insulin possibly by the destruction of β cells in the islets of Langerhans in the pancreas, which secrete this hormone. It is likely that the cells are destroyed by the body's own immune system. This usually starts when someone is quite young.

Type 2 diabetes is an inability of cells to respond to insulin and may be because there are few receptors on the cell surface membranes of target cells (see pages 36 and 73). This form of diabetes is often associated with obesity, a high sugar diet, the inheritance of the alleles of certain genes and ethnicity. People from some ethnic groups, particularly people of Caribbean and South Asian origin, are at high risk of type 2 diabetes. The global increase in diabetes is related to high levels of obesity associated with a change from traditional diets, diminishing levels of physical activity, population ageing and increasing urbanisation. Ninety-five per cent of people with diabetes in the Caribbean have type 2 diabetes. It is estimated that one in 10 of the adult population has it, with levels of one in five for people over 40 in the Americas. It is also increasing in children as levels of obesity increase.

There is no cure for diabetes. Type 1 diabetes is treated by injections of insulin. Type 2 diabetes is controlled by diet and exercise, although insulin injections may be required at later stages. For many years diabetes was treated by regular injections of insulin extracted from animals, such as pigs and cattle, which were slaughtered for the meat trade. This form of insulin is still available although most diabetics now receive human insulin that is prepared from genetically modified bacteria or yeasts.

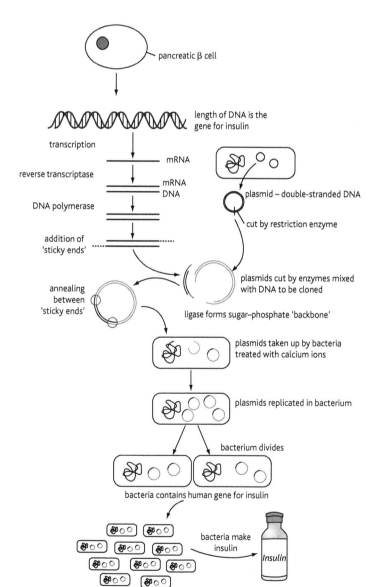

Figure 4.4.1 *This shows the steps involved in producing GM bacteria cells that produce insulin on a commercial scale*

The outline of genetic engineering on pages 112–113 states that genes are cut from DNA. In fact that can be very difficult and the gene used in insulin production was found by isolating the mRNA in pancreatic β cells that make insulin. The mRNA was used as the substrate for the enzyme **reverse transcriptase**, which uses mRNA as a template to make a single-stranded molecule of DNA that has a base sequence complementary to that of RNA. This single-stranded molecule (cDNA) is then replicated to give a double-stranded molecule. It is the double stranded molecule that was inserted into a vector and then into the cells that produce insulin.

Insulin was initially made by genetically engineering bacteria to produce it. However, insulin has two polypeptide chains and bacteria do not modify proteins in the right way to form a functioning insulin molecule. This is why insulin, along with other proteins, is produced in eukaryotic cells, such as yeast cells and animal cancer cells, rather than in bacteria.

Timeline for insulin

Date	Event
1921	insulin discovered by researchers in Toronto
1955	Fred Sanger determined amino acid sequence of insulin
1969	Dorothy Hodgkin used X-ray crystallography to determine the crystalline structure of insulin
1978	US biotechnology firm Genentech synthesised first recombinant insulin in a bacterium, *Escherichia coli* (licensed to the drugs firm Eli Lilly)
1983	pharmaceutical firm Eli Lilly starts selling recombinant insulin as Humulin

Advantages of using GM insulin

The main advantage of this form of insulin is that the amino acid sequence can be changed to alter the properties of the insulin. These are called insulin analogues that either act faster than animal insulin (useful for taking immediately after a meal) or more slowly over a period of between 8 and 24 hours to give the background blood concentration of insulin. Many diabetics take both at the same time. However, studies have not shown any advantage of using them and they are more expensive, and less pure, than animal insulins. The companies that produce animal insulin have withdrawn it from many countries, although it is still available in the UK, for example. The number of people requiring insulin is increasing worldwide. In the Caribbean, diabetes is one of the leading health problems so it is important that there are reliable supplies that are not dependent on factors such as availability through the meat trade.

Disadvantages of using GM insulin

The main disadvantage of GM human insulin is that some people have reported that they do not experience any warning signs of a hypoglycaemic attack when the blood glucose concentration falls increasing the likelihood of a diabetic coma.

∞ *Link*

Reverse transcriptase is an enzyme from viruses that they use to copy their RNA into DNA. See more on transcription on page 68.

Summary questions

1 State why some people require regular injections of insulin.

2 Describe in bullet points the process of modifying cells to produce insulin.

3 Discuss the advantages and disadvantages of using genetically modified cells to produce insulin.

4 Explain the use of the enzyme reverse transcriptase in genetic engineering.

☑ *Study focus*

Weeds are plants that compete with our crops for space, light, water and ions from the soil. Huge amounts of money are spent every year on herbicides to control weeds.

Did you know?

Transgenic papayas carry genes that confer resistance to *Papaya ring spot virus*, which devastated Jamaica's crop in the mid-1990s.

Genetically modified organisms (GMOs) are transgenic organisms. They have one or more foreign genes which are expressed. Promoter sequences of DNA are inserted alongside the gene so that it is expressed in the host organism.

GMOs in agriculture

Humans have always improved crop plants by using artificial selection. GM technology allows the transfer of individual genes between unrelated species. Almost all GMOs that have been released into the environment for testing or for production are crop plants.

GM crop plants

The first GM crop plants to be developed had genes to improve cultivation by incorporating pest and herbicide resistance. Pest resistance reduces losses to insect pests such as the cotton boll weevil; herbicide resistance allows farmers to spray herbicides to kill weeds without killing the crop.

Most recently, crops have been engineered to improve human health. An example is a GM rice known as Golden rice™. Another is a GM banana that will act as a vaccine for hepatitis B.

Feature	Example of crop	Feature that has been added
disease resistance	papaya	resistance to ring spot virus
pest resistance	cotton	toxin coded for by a gene from *Bacillus thuringiensis* to kill pests such as boll weevil
herbicide resistance	soya corn (maize) cotton	gene that gives resistance to effects of herbicide glyphosate so allowing weed control by spraying during growth of crop
drought resistance	corn (maize) sugar cane	genes to control water vapour loss
improved nutritional qualities	rice	several genes to produce precursors of vitamin A in endosperm
	Irish potato	increased starch content

Most GM crops are grown in USA, Brazil and Argentina. Much research on GMOs in crop plants is carried out in Puerto Rico by one of the leading firms, Monsanto. Few GM crops are grown in the Caribbean, although field trials on genetically modified papayas began in Jamaica in 1998.

GM livestock

The first GM animals were developed to improve their growth rate; these animals may enter the food market and the human food chain. More recently, animals have been transformed to produce specific substances in milk, eggs or blood. No animals have been given approval by regulatory bodies for production and consumption by humans (as of 2011).

Feature	Example of animal	Feature that has been added
increase in growth rate	salmon	gene from the fish species ocean pout that gives salmon ability to grow all year round
decrease in pollution through reduction in dietary supplements	pigs ('enviropigs')	enzyme phytase to digest phytins so decreasing need for phosphorus supplements and reducing phosphate pollution
prevent transmission of disease	chickens	catch but do not transmit bird 'flu
production of human proteins in milk	dairy cattle	genes for human lysozyme and other proteins found in human milk

GM microorganisms

Microorganisms were the first organisms to be genetically engineered and there are now many producing a wide range of chemicals often in controlled conditions in laboratories or factories, not in the wider environment.

Feature	Example of microorganism	Feature that has been added
production of animal protein	bacteria	chymosin (rennin) for cheese making somatotrophin for injecting into cattle to improve milk production
wine production	yeast	improving the taste and colour stability of wine as well as to avoid the production of undesirable compounds (histamines)
production of human proteins for medical use	yeast	insulin, human growth hormone, vaccines
dietary supplements	bacteria	tryptophan (an amino acid)

Genetically-modified organisms in medicine

GMOs are used to produce medicines that are difficult to produce in other ways. This is a list of some of the many products for treatment and medical research:

- insulin production (see page 116)
- human growth hormone
- thyroid stimulating hormone
- factor VIII – a blood clotting protein
- vaccines, e.g. for influenza
- many chemicals for research including monoclonal antibodies

Did you know?

In a year each GM goat produces as much antithrombin as can be collected from 90 000 blood donations. (See Question 4 on page 122.)

Did you know?

Antithrombin inhibits thrombin, an enzyme that promotes blood clotting. Antitrypsin inhibits the protease trypsin, which is released by phagocytes in the lungs during infections. These recombinant drugs are used to treat people who have inherited deficiencies of these anti-enzymes.

✔ *Study focus*

Many children worldwide do not receive enough of this vitamin in their diet. It is estimated that 140 to 250 million children under the age of five suffer from vitamin A deficiency. Experts estimate that half a million children have such severe deficiency that they have become blind.

✔ *Study focus*

Read more about GM technology online, but take note that if you read websites you should check which side of the GM debate they advocate and treat their evidence and opinions accordingly.

- human antithrombin (ATryn®) to treat blood loss produced in milk by transgenic goats
- human alpha-antitrypsin (ATT) to treat emphysema produced in milk by transgenic sheep.

The advantage of using these organisms is the large production and therefore cheaper prices of the substances concerned.

Some of the precautions to contain GM microorganisms in laboratories and in factories to prevent their release into the environment are:

- transgenic microorganisms do not compete well in the natural environment as they are engineered to produce substances that give them no advantage; this requires energy that is not readily available in the 'wild' to produce these substances
- containment facilities, such as filters on air conditioning and air locks on doors, prevent escape of organisms
- lethal genes are added to the microorganisms so that they die if removed from the conditions of the culture.

Golden Rice™

The problem: Vitamin A deficiency

The following gives you some more information about the genetic modification of rice.

Vitamin A is required in the diet for the health of epithelia and the retina in the eye.

Vitamin A deficiency may result in:

- failure of rod cells in the retina to make the pigment necessary to see in low light intensity (night blindness)
- dry, ulcerated cornea that becomes cloudy, leading to blindness
- poor repair of epithelia leading to increased risk of infection, especially measles

In the Caribbean, children in Haiti are at highest risk of vitamin A deficiency.

Solutions

- provide vitamin A supplements, for example every four to six months; providing these at the same time as giving vaccinations for polio and measles has been very successful
- encourage mothers to breast feed and provide them with vitamin A supplements to increase content of breast milk
- add vitamin A to foods, such as cooking oil, flour, milk and sugar; these are called fortified foods.

Governments and organisations, such as UNICEF, have supplementary feeding programmes in many parts of the world. This is effective if children receive doses of vitamin A early enough and at regular intervals. Many fortified foods do not reach the poorest who often live beyond the reach of health services, voluntary and aid organisations. Many of these children exist on a very poor diet consisting mainly of a staple food, such as rice or maize and little meat or vegetables.

We are able to metabolise the yellow pigment, β-carotene, into vitamin A. β-carotene is in many plants, including rice, but not in the part of the rice grain that most people eat. Plants have genes that code for the enzymes that make β-carotene from other substances. White rice is the endosperm of the grains and although the genes that code for these are in the cells, they are not transcribed and translated.

Substances in the pathway	Enzyme	Genes inserted into endosperm	Source of genes
precursor			
↓	←— 1	*psy*	maize
B			
↓	←— 2		
C		*crt1*	bacterium *Erwinia uredovora*
↓	←— 2		
D			enzyme 3 is expressed naturally in the endosperm
↓	←— 3		
β-carotene			

The solution to this problem was to genetically modify rice plants using genes taken from maize and a bacterium. Researchers in the Golden Rice™ project used the bacterium, *Agrobacterium tumefaciens*, to infect the embryos of rice plants. This bacterium has a tumour inducing (T$_i$) plasmid that moves from the bacterium to the host cells and becomes incorporated into the rice chromosomes. Genes coding for the enzymes to make β-carotenes were inserted into these plasmids. The embryos grew into plants which were self-pollinated so the new genes were inherited by future generations. They can be incorporated into different varieties of rice by cross breeding.

Field trials have been carried out in the USA and the Philippines. It is possible that Golden Rice™ will become available to farmers in the next few years. At the time of writing (2011) it is not. The project has met with much resistance; opponents maintain that it is better to widen the diet of those at risk of vitamin A deficiency, for example by increasing intake of leafy vegetables.

Moral and ethical issues of GMOs

There are many objections that people make to the development, use and release of GM organisms. Some of these are:

- Antibiotic resistance genes used to identify GMOs could 'escape' and be transferred to pathogenic organisms making some antibiotics redundant.
- Herbicide resistance genes could be transferred in pollen to weed species and lead to the development of 'superweeds' that are resistant to herbicides.
- Foreign genes could be transferred to wild relatives of our crop plants so changing their genomes; this may 'pollute' those species that may prove useful sources of genes for crop improvement in the future.
- Foreign genes can 'pollute' non-GM and organic crops, which require certification that they provide 'GM-free' food.
- GMO crops require more herbicide applications and just as much pesticide as non-GM crops so there is no advantage in terms of cost to farmers or reduction in chemicals used in agriculture.
- Farmers cannot keep seed for sowing for the following crop as GM crops do not 'breed true'; this favours large-scale commercial farmers and not many farmers in developing countries.

A. tumefaciens and its plasmid infect plants to cause crown gall which is a cancerous disease. Genetic engineers use the bacterium as a delivery system to get the plasmid into plant cells. The researchers used rice embryos because the bacterium does not infect adult cereal plants, such as rice.

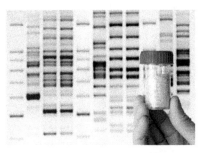

Figure 4.4.1 Some Golden Rice™ in front of some DNA profiles. β-carotene is a yellow pigment, hence the name 'Golden Rice'.

 Study focus

You should read more about GM technology and then discuss the issues raised by this example and others with your friends, family and teachers. The discussions will help you form your own opinions of this technology and provide evidence to support them.

Summary questions

1 Find out about: **i** the uses of GM soya and GM maize; **ii** GM microorganisms. Present your findings as a poster or a group presentation.

2 Read objections to genetically modified food. Make a poster outlining the benefits of the use of GMOs in agriculture and the objections to their use.

3 Explain the advantages of using GMOs in the production of medicines.

Find below a selection of short answer questions on genetic engineering. Some questions require information from earlier chapters.

1 **a** Explain how the following enzymes are used in genetic engineering: restriction enzymes, reverse transcriptase and ligase.

   b Explain what is meant by the term *vector* and explain how different vectors are used in the genetic modification of bacteria.

   c Outline the roles of bacteria in genetic engineering.

2 The DNA target sites (restriction sites) for three restriction enzymes are shown below. The vertical lines indicate where the enzymes cut DNA.

Hindlll	5'...A|AGCTT...3' 3'...TTCGA|A...5'
EcoRI	5'...G|AATTC...3' 3'...CTTAA|G...5'
Haelll	5'...GG|CC...3' 3'...CC|GG...5'

   a **i** Describe the features of the restriction sites shown in the diagram.

   ii With reference to the information in the diagram, explain the advantages of using restriction enzymes in genetic engineering.

Three identical lengths of DNA were treated *separately* with each of the restriction enzymes. The DNA has the following sequence of base pairs.

5' AGTTGAAAGGCCTTCATCGCACCCTTAATTCGTGGCCAAGCTT 3'
3' TCAACTTTCCGGAAGTAGCGTGGGAATTAAGCACCGGTTCGAA 5'

   b **i** State how many fragments of DNA will be present after treatment with EACH of the restriction enzymes.

   ii Explain why some of the fragments need further treatment before they can be inserted into plasmid vectors.

Retroviruses have RNA as their genetic material.

   c RNA can be incorporated into the genome of retroviruses for insertion into bacteria or eukaryotic cells. Explain how this technique is used to insert genes into the genome of these transformed cells.

3 The first technique to produce human insulin by genetic engineering involved inserting a gene for human insulin into the DNA of a plasmid.

   a Name the enzyme which would be used to **i** cut the plasmid DNA and **ii** insert the DNA for human insulin into the cut plasmid.

There are 51 amino acids in insulin, made up of 16 of the 20 amino acids that are coded for by DNA.

   b What is the minimum number of different types of tRNA molecules necessary for the synthesis of insulin? Explain your answer.

   c Explain why the number you have given in **b** is likely to be fewer than the actual number of different types of tRNA required.

   d A triplet in the template strand of the gene for insulin is TAG. Explain the likely effects of a mutation in the first base from T to A.

4 The diagram shows how the gene for human antithrombin was introduced into goats.

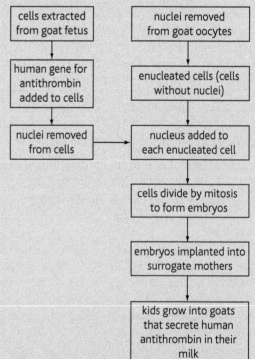

   a Explain what is meant by a transgenic animal.

   b Complex human proteins such as antithrombin cannot be produced by genetically modified bacteria. Explain why bacteria are unable to produce complex human proteins such as antithrombin.

Animal cells in culture have also been modified to make human proteins, such as insulin and human growth hormone.

c i Discuss the advantages of genetically modifying animals, such as goats, and animal cells, such as Chinese hamster ovary cells, to produce human proteins.

ii Outline the potential hazards in using genetic engineering to produce human proteins.

5 a State what is meant by the term *gene therapy*.

b Explain, in outline, how gene therapy is carried out.

c Outline the advantages and hazards of gene therapy.

d What are the limits of gene therapy?

6 Several crop plants have been genetically engineered to express poisons based on a toxic compound from the bacterium, *Bacillus thuringiensis* (Bt). GM varieties of soya, oil seed rape, cotton, maize and tobacco are all grown in countries such as the United States, China and Brazil.

a Explain the advantages of growing Bt varieties of these crop plants.

b Outline how crop plants, such as those listed above, are genetically modified to improve productivity.

c Outline the potential risks of growing GM crop varieties and suggest steps that can be taken to minimise them.

7 Mexico banned the cultivation of GM maize in 1998. In an investigation carried out in 2000 researchers tested samples of maize grown in remote mountain regions of Oaxaca in southern Mexico for a specific sequence of DNA that is only found in GM maize. The researchers tested for a promoter sequence that genetic engineers use in maize; the sequence originates from a virus that infects cauliflowers. They also tested blue maize from the remote Cuzco valley in Peru and 30-year-old maize cobs from museum collections in Mexico. The table shows their results.

Sample of maize	Presence (✓) or absence (✗) of promoter sequence in maize genome
local variety of maize 1	✓ (1–3% of grains sampled)
local variety of maize 2	✓ (1–3% of grains sampled)
GM Bt maize from USA	✓ (100% of grains sampled)
GM herbicide resistant maize from USA	✓ (100% of grains sampled)
blue maize from Peru	✗
maize from 30-year-old museum collection	✗

a Explain why the researchers used **i** blue maize and 30 year old maize, and **ii** GM maize from the USA in their study.

b Suggest reasons for the presence of the promoter sequence in the genome of maize grains from southern Mexico two years after the ban on GM maize.

This research was highly controversial and repeat investigations in 2003–04 did not find any of these DNA sequences in Mexican maize. A third investigation in 2008 found some, but did not confirm the original conclusion that the DNA had entered the germ line.

In 2009, the Mexican government permitted trial plantings of GM maize.

c Explain what is meant by the term *germ line*.

d Discuss the reasons why:

i some countries permit the planting of GM crops

ii others do not permit the planting of GM crops.

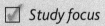 *Study focus*

Promoter sequences are attachment sites for RNA polymerase and transcription factors. The sequences are 'upstream' of each gene. Any gene inserted by genetic engineering must have a promoter sequence to allow it to be transcribed in the host cell.

5.1 Variation (1)

☑ Study focus

During your course you will probably do some field work and use an identification key in the form of a field guide.

∞ Link

Read page 90 for the differences between phenotype and genotype.

☑ Study focus

The term *population* will be used often in this section. It is all the organisms of one species in the same area at the same time.

Types of variation

Variation refers to differences between the genotypes and phenotypes of organisms.

Phenotypic variation is obvious to us in the physical appearance of organisms and in their behaviour. It is also evident in aspects of biology that we cannot see so easily, such as biochemistry and physiology. Examples of this are the variation in haemoglobin (see page 130) and blood group antigens (see page 102). Some phenotypic variation is a direct reflection of genotypic variation as it is with the codominance in flower colour of the 4 o'clock plant (see page 96).

We can identify phenotypic variation at two levels:

- **interspecific variation**
- **intraspecific variation**.

Interspecific variation is the differences *between* different species. We use these differences to distinguish between different species; biologists use these differences when classifying organisms and when using keys to identify species.

In this section you will need to know about intraspecific variation.

Intraspecific variation is variation *within* a species. It is the difference between individuals. There are two types of intraspecific variation:

- **discontinuous variation**
- **continuous variation**.

Discontinuous variation is the type of variation in which there are clear, non-overlapping categories. You can see examples of this type of variation in humans in the list on page 90. People either have attached or non-attached ear lobes. There are two distinct categories with no intermediate forms. Other features in the list, such as height and hand span, show continuous variation as there is a range of sizes between two extremes.

Discontinuous variation

This refers to qualitative differences that are in clear contrasting categories. Fruit flies either have long wings or short wings, there are no intermediate wing sizes; blood groups in humans show discontinuous variation – in the ABO system there are four categories – A, B, AB and O – without any intermediates between them.

This type of variation is caused by genes; the environment has no effect. For example, the ability to taste a bitter chemical, PTC, is determined by a gene (TAS2R38) on chromosome 7. The dominant allele gives people the ability to taste PTC; people who are homozygous recessive cannot taste it. The types of food we eat may modify how we taste PTC, but cannot make non-tasters into tasters.

Data on discontinuous variation is presented in the form of bar charts because each category is a separate group.

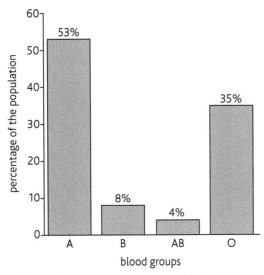

Figure 5.1.1 *Bar chart to show percentages of the population of Portugal with different ABO blood groups*

Rules for drawing bar charts:

- use most of the grid provided, do not make the chart too small
- draw the chart in pencil
- bar charts can be made of lines, or more usually, blocks of equal width; there must be space between the lines or bars; they do not touch
- the intervals between the blocks on the *x*-axis should be equidistant
- the *y*-axis should be properly scaled with equidistant intervals; the scale should usually start at zero and this should be written at the base of the axis; if all the numbers are large a displaced origin may be used but the start number should be clear at the base of the *y*-axis
- the *y*-axis should be labelled with the headings and units taken from the table of results
- the lines or blocks should be arranged in the same order as in a table of results or in a logical sequence from left to right (increasing or decreasing)
- each block should be identified clearly.

Summary questions

1 Define the term *variation*.

2 Distinguish between the following pairs: interspecific and intraspecific variation; discontinuous and continuous variation.

3 The proportion of different blood groups in Brazil is: A 41%, B 9%, AB 3%, O 47%.
   a Plot these figures as a bar chart.
   b Brazil was a Portuguese colony until 1822 and many Portuguese migrated there. Suggest why these percentages are different from those in Portugal.

4 Describe why variation in blood groups is an example of intraspecific variation and discontinuous variation.

5 State why the proportion of a population with different blood groups is plotted as a bar chart.

6 Explain the steps you would take in determining and presenting the variation in the ability of people to taste a bitter chemical called PTC.

5.2 Variation (2)

Learning outcomes

On completion of this section, you should be able to:

- list examples of continuous variation
- select and draw histograms to present data on continuous variation
- describe some effects of the environment on the phenotype.

☑ Study focus

The type of variation shown in the histogram below is called a normal distribution as the three measures of the 'average' (mean, median and mode) are all about the same.

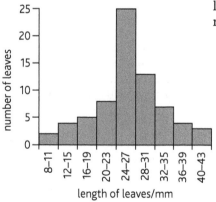

Figure 5.2.1 *Histogram to show the variation in leaves recorded in the tally table*

Continuous variation

This refers to quantitative differences within a species.

Organism	Examples of continuous variation
humans	height, body mass, hand span,
dairy cattle, goats	milk yield, e.g. volume of milk per week
beef cattle, pigs, sheep, goats	body mass, carcass weight
sheep	wool thickness
soya	width of leaves, mass of pods, total yield of beans
banana	length of leaves, yield of fruit

Each of these examples has no clear categories; instead, there is a range between two extremes – shortest and tallest, lightest and heaviest, etc.

To describe continuous variation you need to measure and then collate the data. The results are usually collected in a tally table before being presented as a histogram. In collating the results it is necessary to divide the range into classes. For example, if you were collecting information on lengths of 50 leaves with a total range of 8 to 38 mm, you could divide the range into nine classes. This is shown in the tally table and histogram.

Leaf length/mm	Tally	Number	Percentage frequency
8–11	//	2	3
12–15	////	4	6
16–19	⊁⊬⊬	5	7
20–23	⊁⊬⊬ ///	8	11
24–27	⊁⊬⊬ ⊁⊬⊬ ⊁⊬⊬ ⊁⊬⊬ ⊁⊬⊬	25	35
28–31	⊁⊬⊬ ⊁⊬⊬ ///	13	18
32–35	⊁⊬⊬ //	7	10
36–39	////	4	6
40–43	///	3	4
Total	71	71	100

Continuous variation is determined by genes and by the environment. Many genes influence the same feature which is known as **polygeny**. Often the genes involved have multiple alleles so that many alleles at each locus influence the feature. The separate genes and the alleles at each locus influence the feature in an additive fashion. In a simple example with just two genes, each recessive allele could add 5 mm to the length of a leaf and each dominant allele add 10 mm. Leaves would then vary in length between 20 mm and 40 mm so there would be five different classes. However, features such as leaf length are influenced by many more than two genes and environmental effects, such as light, water and nutrient availability, have the effect of smoothing out the variation to give a **bell-shaped curve** if we draw a line through the centre of the bars on the histogram.

Histograms are used to present continuous variation as a continuous scale between extremes; the items are not separate categories.

Rules for drawing histograms:

- use most of the grid provided, do not make the histogram too small
- draw the histogram in pencil
- the *x*-axis, represents the independent variable and is continuous; it should be labelled clearly with an appropriate scale
- the blocks should be drawn touching
- the *area* of each block is proportional to the size of the class; the leaf-length data in the table opposite is in similar-sized classes so the widths of the blocks are all the same
- the blocks should be labelled, e.g. '3.0 to 3.9' which means that 3.0 is included in this class, but 4.0 is not; 4.0 will be included in the next class: 4.0 to 4.9
- the *y*-axis represents the number or frequency and should be properly scaled with equidistant intervals; label with appropriate units.

This table campares discontinuous and continuous variation.

	Discontinuous variation	**Continuous variation**
appearance of phenotype	qualitative, e.g. presence or absence of a feature with no intermediates	quantitative, e.g. mass and length; a range with many intermediates
number of genes influencing the feature	few, often just one gene (monogenic)	many (polygenic)
effects of different genes on the feature	different effects	all have the same effect, usually additive
effects of alleles at each locus	large effects	small effects, usually additive
effect of the environment	small or non-existent	large
presentation	bar chart	frequency histogram

Effect of the environment on phenotype

An extreme case of the effect of the environment on phenotype occurs in honey bee colonies. Most female honey bees develop into worker bees because they are fed on a diet of pollen and nectar. Young female bees destined to become princesses and then queens are fed royal jelly.

The effects of the environment are not inheritable as changes that have occurred to an organism during its lifetime are not transmitted to the genes in the organism's reproductive organs.

Summary questions

1 Define the term *continuous variation*.

2 Distinguish between the following pairs: bar chart and histogram; monogenic and polygenic.

3 Explain the steps you would take in measuring and presenting variation in pod length of Royal Poinciana, *Delonix regia*.

4 Suggest how a female Siamese cat has black markings, but none of her kittens have them. See page 73 for ideas.

5 Discuss the effects of the environment on the phenotype of individual organisms and explain why these effects are not inheritable.

5.3 Mutation

Learning outcomes

On completion of this section, you should be able to:

- explain the term *mutation*
- describe gene mutation and give an example
- discuss chromosome mutation and give an example.

☑ Study focus

If a mutation occurs in a germ line cell, then the cells derived from it will divide by meiosis. 50% of the haploid cells produced by meiosis will carry the mutant allele.

The term mutation derives from a word meaning change and it is used in biology to refer to changes in DNA. A change occurs in a single cell and is passed to all the descendants of that cell by replication and mitosis.

Proto-oncogenes and tumour suppressor genes control mitosis. Mutations of these genes disrupt this control so that cells divide uncontrollably giving rise to cancers. Mutations that only affect the organism in which they occur and are not passed on to the next generation are known as **somatic mutations**.

Even more important for this chapter are **germ line mutations** that occur in gamete-forming cells and can be passed on to the next generation.

Changes may occur that affect whole chromosomes or single genes.

- **Chromosome mutations** occur when a nucleus divides by mitosis or meiosis; these can either change the number of chromosomes in a nucleus or the structure of individual chromosomes.
- **Gene mutations** occur during DNA replication during interphase of the cell cycle; these are changes to the number or sequence of base pairs.

Chromosome mutation

These may involve:

- **polyploidy** – increasing the number of sets of chromosomes
- **aneuploidy** – increasing or decreasing the number of individual chromosomes
- changes in the structure of a chromosome, such as a **translocation** when part of a chromosome is detached and reattached to another.

Down's syndrome

Down's syndrome is an example of aneuploidy. In most cases it happens because a child has an extra chromosome 21. This happens because homologous chromosomes do not separate properly in anaphase I of meiosis. This occurs most often in the production of eggs rather than in the production of sperm. The reason is that cells may take over 40 years to proceed from early meiosis and go through anaphase I (see page 83) to form eggs. The incidence of Down's syndrome is higher among the children of older women. The chances rise steeply after the age of 35.

Figure 5.3.1 shows how a pair of chromosomes fails to separate in anaphase I and are inherited together. The existence of three chromosomes of the same type is known as **trisomy**. Trisomies are not inherited; they arise because of non-disjunction during meiosis to form gametes. Females with Down's syndrome are fertile and have had children both with and without Down's. Males with Down's syndrome are sterile. Trisomies of other chromosomes occur, but only babies with trisomies of the X chromosome and No. 21 survive. Trisomies of other

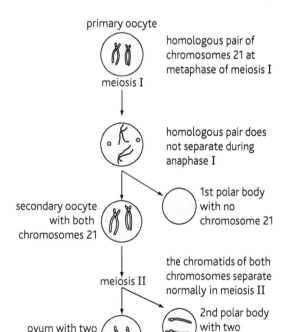

Figure 5.3.1 *This diagram shows how a pair of homologous chromosomes fails to separate during meiosis I*

primary oocyte

homologous pair of chromosomes 21 at metaphase of meiosis I

meiosis I

homologous pair does not separate during anaphase I

1st polar body with no chromosome 21

secondary oocyte with both chromosomes 21

the chromatids of both chromosomes separate normally in meiosis II

meiosis II

2nd polar body with two chromosomes 21

ovum with two chromosomes 21

fertilisation

normal sperm with one chromosome 21

zygote with trisomy 21

chromosomes are usually lethal before or around the time of birth. About 5% of cases of Down's syndrome are the result of translocation that can be inherited.

Changes in the total number of chromosomes have been important in the evolution of plant species, especially in those that are cultivated. Nearly 50% of flowering plant species are polyploid. Polyploidy is often associated with large size, so cultivated varieties are often much larger than the diploid wild relatives from which they evolved.

Gene mutation

Changes to individual genes involve the sequence of nucleotide bases. These can be:

- **substitution** – the change of one base pair for another
- **frameshift** – in which the sequence of triplets changes by the addition of one or more bases or the deletion of one or more bases
- **stutter** – these are mutations which increase the number of repeat sequences of base pairs. The gene for a protein in the nervous system has a repeated section and this is duplicated so that it becomes longer with each generation. This is the cause of Huntington's disorder (see page 109).

Figure 5.3.2 shows the effect of the changes on a sequence of bases in DNA and the amino acid sequence for which it codes. The upper strand is the template strand, the lower is the coding strand.

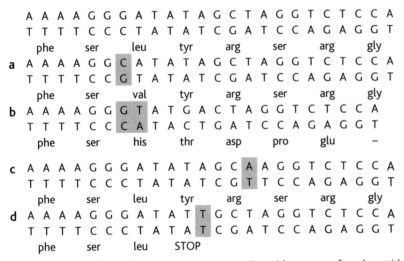

*Figure 5.3.2 The effects of gene mutations on an amino acid sequence of a polypeptide can be: **a** quite small; **b** very large; **c** non-existent; **d** catastrophic*

The chances of a mutation are increased by exposure to **mutagens**, which are environmental factors that interact with DNA to change it. Examples are radiation (X-rays) and various chemicals such as benzpyrene in tobacco smoke. Agents that cause mutations that cause cancers are known as **carcinogens**.

DNA polymerase has a proof-reading capacity so removes and replaces a nucleotide that is in the wrong place. However, this proof reading ability is not infallible and errors occur during replication.

Did you know?

Polyploidy is uncommon in animals; some species of lizard are an example.

∞ Link

Polyploidy is a form of very abrupt speciation – see page 139 for details of how this happened in cord grass.

∞ Link

Look back to the details of the genetic code on page 67 to help you to understand the effects of mutation.

Summary questions

1 Define the terms *mutation* and *mutagen*.

2 Distinguish between *gene mutations* and *chromosome mutations*.

3 Find out the effects of trisomy 21 on the phenotype. Present the information as a table showing the effects on different aspects of he phenotype.

∞ Link

To remind yourself about R groups of amino acids, see page 12.

✓ Study focus

An example of a substitution mutation is shown in Figure 5.3.2a on page 129.

Did you know?

In West Africa the incidence of sickle cell anaemia is 4% of all live births.

A substitution mutation

Sickle cell anaemia is the result of a substitution mutation in the sixth triplet of the gene for β-globin of haemoglobin. This does not sound like a very important change, but as you can see in Figure 5.4.1 it can have very severe consequences if *all* the β-globin polypeptides in haemoglobin are changed. The figure shows that the change from A to T in the sixth triplet of the coding strand leads to the change from CTC to CAC in the template strand and hence the substitution of valine for glutamic acid in the sixth position of the primary structure of β-globin. This type of haemoglobin is known as haemoglobin S (HbS) to distinguish it from the normal form (HbA).

Molecules of haemoglobin change shape when they gain and lose oxygen. The form with no oxygen attached exposes an area on its surface that has R groups of phenylalanine and leucine that is hydrophobic. The hydrophobic R group of valine at position 6 of the beta globin is attracted to this hydrophobic area, causing haemoglobin S molecules to stick together and precipitate.

When cells with the mutant polypeptide enter regions of low oxygen concentration they become sickle-shaped or crescent-shaped. The repeated sickling and unsickling damages the red cell membrane and shortens the life of these cells. They are removed from the circulation by the spleen. These sickle cells can get stuck in capillaries making it difficult for blood to flow. If this happens, people with this condition have sickle cell crises in which blood flow to major organs is reduced and there can be much pain. There can also be fever and difficulty in breathing. Blood flow to the spleen may increase so much that there is not enough blood for the rest of the body. People experiencing a sickle cell crisis need to be taken to hospital to receive pain relief and blood transfusions. Some people live with the disease for many years without having crises, but without good medical facilities many children with sickle cell disease die at a young age.

The anaemia is caused by the loss of red blood cells at a faster rate than they are produced in bone marrow. Normal red blood cells live between 90 and 120 days before they are destroyed in the spleen or liver; sickle-shaped red blood cells only live for between 10 and 20 days. Symptoms of anaemia are feeling tired and lethargic and an inability to do strenuous exercise.

Figure 5.4.1 shows the change in DNA and the cascade of effects that this has on the body in someone who is homozygous for the mutant gene. Notice that a gene can have multiple effects on cells, tissues, organs and organ systems. **Pleiotropy** is the term applied to genes like the β-globin gene that has multiple effects on the phenotype.

Protection against malaria

Sickle cell anaemia is the disorder that happens to people homozygous for the substitution mutation. It is surprisingly common in some areas of the world because it gives protection against malaria. The malarial parasite, *Plasmodium*, has a complex life cycle and spends part of it in red blood cells. Invasion by the parasite causes the oxygen concentration within the red cells to decrease, which causes the red cells to become misshapen. They are then taken out of the circulation and killed. Not only does this kill the cells but also the parasites within. Unfortunately, this protection against malaria has little benefit for those with the disorder since sickling occurs anyway and, without medical intervention, the condition is lethal.

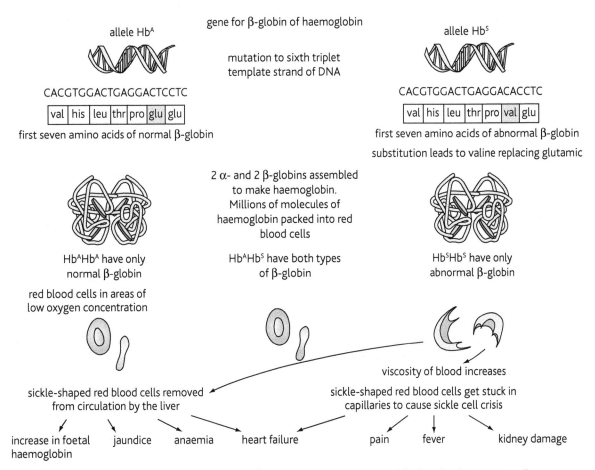

Figure 5.4.1 *The effects of a substitution mutation in the β-globin gene. Viscosity means 'thickness' and resistance to flow.*

Both the normal allele and the mutant allele are active in people who are heterozygous. They produce both forms of the polypeptide and their blood functions fairly normally as they have haemoglobin molecules with two normal β-globins and some with two mutant β-globins and some with both types. In a carrier, the presence of the malarial parasite causes the red blood cells to rupture prematurely, so the parasite cannot reproduce. Furthermore, the polymerisation of haemoglobin affects the ability of the parasite to digest haemoglobin. Therefore, in areas with malaria, people's chances of survival actually increase if they have both forms of the β-globin in their haemoglobin. This mild form is known as sickle cell trait and is common in places such as West and East Africa as a result of natural selection (see page 134).

People inherit sickle cell anaemia from parents who are carriers. People who are heterozygous and carry one mutant allele are often unaware that they are like this. Some carriers do report mild symptoms. Alpha and beta thalassaemia are other genetic disorders of haemoglobin. The mild forms of the thalassaemias also provide protection against malaria. Sickle cell anaemia and the serious forms of thalassaemias are serious health problems in areas of the world where malaria is (or was) an important infectious disease. They are also important in communities that are descended from migrants from Africa and from the Mediterranean area, where thalassaemias are common.

Summary questions

1 The gene **Hb^A**/**Hb^S** controls the production of the β-globin polypeptide of haemoglobin. People who are homozygous, **Hb^S Hb^S**, have sickle cell anaemia. Draw a genetic diagram to show how someone can inherit this disorder from their parents neither of whom has the disorder.

2 Explain the effect of the change to the DNA nucleotide sequence in the gene for β-globin on the structure and function of red blood cells.

3 The terms dominant, recessive and codominant are used to describe the relationships between alleles. What term or terms do you think should be applied to the alleles of the gene for β-globin? Justify your answer.

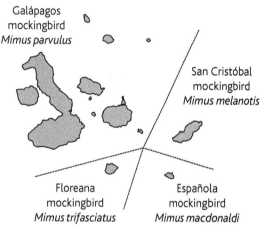

Figure 5.5.1 *This map of the Galápagos Islands shows the distribution of the four species of mockingbirds that are endemic to these islands – found here and nowhere else*

Galápagos mockingbirds

Evolution will forever be linked to the name of Charles Darwin (1809–1882), the English naturalist and scientist whose journey around the world in the ship *HMS Beagle* between 1831 and 1836 led him to propose a mechanism by which evolution occurs.

On his circumnavigation of the globe Darwin visited the Cape Verde islands off the west coast of Africa, South America, islands in the Pacific, New Zealand, Australia and South Africa. While on the Galápagos Islands in the Pacific he made this observation about birds that were then known as mocking thrushes.

'My attention was first thoroughly aroused, by comparing together the numerous specimens .... of the mocking-thrushes, when, to my astonishment, I discovered that all those from Charles Island [now called Floreana] belonged to one species (Mimus trifasciatus); all from Albemarle Island [Isabela] to M. parvulus; and all from James Island [Santiago] and Chatham Island [San Cristóbal] belonged to M. melanotis.'

These birds are now known as mockingbirds and as a group they are widely distributed in the Americas. Darwin had seen them on the South American mainland and realised that the Galápagos birds were related to them. There are in fact four species of *Mimus* in the Galápagos, Darwin only saw three as *M. macdonaldi* is found only on Española – an island Darwin did not visit.

The variation amongst the mockingbirds on the different islands sowed the seed for his big idea: that changes in species occur by **natural selection**.

The close resemblance between the four mockingbird species and species on the mainland led Darwin to suggest that individuals had migrated from the mainland to populate the islands. He then stated that populations of birds isolated on islands with different environments led to changes significant enough to give these populations features that they did not share with mockingbirds on other islands.

Darwin's finches

Darwin collected specimens of finches, which are smaller than mockingbirds, from the islands he visited. These were later studied in London by John Gould (1804–81) who realised that they were closely related species that were significantly different from related species from the mainland. They are now known as Darwin's finches. The beaks of these finches show adaptations to different foods and food gathering methods. The woodpecker finch uses cactus spines to remove insects from the bark of trees; ground finches feed on seeds.

Darwin found other species of animals and plants found nowhere else. The giant tortoise is perhaps the best known. A species that is only found on an island and nowhere else is known as an **island endemic**.

On his return to England in 1836, Darwin spent 20 years collecting evidence for his theory and communicating with scientists, naturalists and animal and plant breeders. He compiled many examples of variation amongst plants and animals.

Natural selection

In 1858, the naturalist and collector Alfred Russel Wallace sent Darwin an essay that included much the same ideas as his own on changes that occur to species. Darwin and Wallace's ideas were presented at a meeting of the Linnean Society in London. In 1859, Darwin's book *On the Origin of Species by Natural Selection* was published.

We can summarise Darwin's observations as follows:

- overpopulation – all species have the ability to reproduce large numbers producing far more individuals than ever become mature; Darwin calculated that a pair of elephants had the potential to have 19 million descendants after 700 years
- populations of organisms remain fairly stable from year to year with minor fluctuations
- within each species there is variation
- offspring resemble their parents.

His theory of natural selection stated that:

- there is competition for resources between individuals of the same species that have the same means to obtain those resources; animals compete for water, food, territories, nesting sites and mates; plants compete for space, water, ions, light, carbon dioxide and, in some cases, pollinators
- competition leads to high death rates among young animals that starve, are eaten by predators or die of disease; there is a similar high mortality among seedling plants that start growing in unsuitable places, do not absorb enough water, ions, light or carbon dioxide, are eaten by herbivores or are killed by disease
- in this struggle for existence the individuals best adapted to obtain resources survive to breed and pass on their alleles
- this **differential survival** means that populations consist of those individuals best adapted to the conditions existing at any one time
- the best adapted animals are good at finding food, escaping from predators, resisting disease and finding mates.

Darwin could not explain how variation was inherited. If he had read Mendel's paper and understood it, he might have made an even more convincing case for natural selection. As it was he thought that 'blending' inheritance occurred and this did not provide the support for natural selection that knowledge of Mendelian genetics would have done.

Did you know?

Darwin was particularly interested in the variation that occurs amongst domesticated animals, such as pigeons, and spent time talking to pigeon breeders and even breeding his own pigeons.

Figure 5.5.2 *A Galápagos mockingbird,* Mimus parvulus

☑ Study focus

Competition between members of the same species is intraspecific competition – a useful term to use in an answer on natural selection.

☑ Study focus

Notice the term *alleles* is used and not genes. All individuals of the same species have the same genes – they differ in the alleles of those genes. Some alleles give individuals the competitive edge that increases their chance of survival.

Summary questions

1 Define the term *island endemic*.

2 Anolis lizards are found throughout the Caribbean. Suggest why Anolis lizard species from St. Lucia are different to those from Jamaica.

3 Explain the importance of the mockingbirds of the Galapágos Islands in the development of Darwin's theory of natural selection.

4 State Darwin's four observations about natural populations and explain the significance of each one for the theory of natural selection.

5 Suggest why Darwin was interested in the work of animal and plant breeders.

Did you know?

Melanism is not only associated with pollution. Insects living in colder environments are often melanic so they absorb heat, warm up and become active faster than non-melanic forms. There are melanic bananaquits on Grenada and St Vincent (see page 140).

∞ Link

See page 140 for experimental evidence relating to the effect of selection on peppered moths.

*Figure 5.6.1 Speckled and melanic peppered moths settle on trees: **a** in an area free of air pollution and covered with lichen; **b** in an area heavily polluted with sulphur dioxide and soot*

Evolution – the big idea

When people talk about evolution they are usually referring to two distinct ideas:

- The general theory of evolution states that organisms change over time. As scientists learnt more about the Earth in the eighteenth century there came a general realisation that the Earth was very old and that life had changed over time.
- The special theory that evolution occurs by the process of natural selection.

Darwin proposed a mechanism by which change in species could occur. He did not know how the variation that he described at great length was inherited. From 1900, biologists have combined knowledge of genetics with ideas of natural selection. Many investigations of natural selection have been published including:

- **industrial melanism** in the peppered moth, *Biston betularia*
- **antibiotic resistance** in bacteria.

Natural selection in action

The peppered moth and industrial melanism

The peppered moth, *Biston betularia*, is found throughout the British Isles. It flies at night and during the day settles on trees where it is camouflaged against lichens that grow on bark in unpolluted parts of the country. There are two very distinct forms within this species. One has fairly light coloured wings speckled with black that makes it look as if its white wings have been dusted with black pepper. This form is well camouflaged against the lichens that grow on trees. The other form is black and is known as the melanic form.

As soon as any melanic moths appeared they were likely to be eaten by predatory birds since they were so obvious against the speckled background on trees. They were eaten before they had the chance to reproduce. These melanics re-appeared now and again by mutation, but the mutant allele was not inherited.

In the middle of the 18th century, the environment changed significantly. With the industrial revolution came severe air pollution from the burning of coal. Smoke contains sulphur dioxide, which killed most of the lichen species, and soot, which was deposited on the trees. In woodlands around Manchester people noticed more and more of the melanic peppered moths which were now camouflaged against the trees. The speckled moths were easily spotted by birds and eaten while the melanics survived and reproduced. The melanics left more offspring than the speckled variety so that by the beginning of the 20th century the melanic form made up over 90% of the population in woodlands around industrial cities. The composition of populations in rural areas with no pollution has not changed, although in eastern England where there is little pollution the melanic variety increased as pollution was carried by the prevailing wind.

The map shows the situation in the mid 1900s. Now, with the decline of heavy industry in the UK and the introduction of Clean Air Acts there is much less pollution and the population has changed with the speckled form increasing in numbers and the numbers of melanics decreasing.

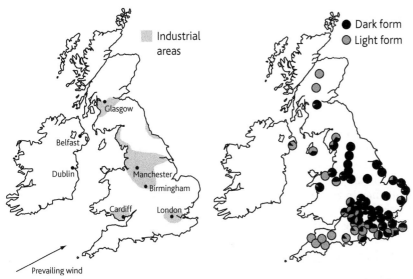

Figure 5.6.2 *These maps shows the proportion of populations of peppered moth in the British Isles that were melanics in the mid-1900s*

In this example, bird predators are acting as the agent of natural selection although their effect was due to pollution in the environment as a consequence of human activity.

Antibiotic resistance

Antibiotics are substances produced by fungi and some bacteria that kill or inhibit the reproduction of bacteria. Antibiotics were developed as medicines in the 1940s with penicillin and streptomycin being the first. Penicillin works by inhibiting the formation of cell walls. Streptomycin inhibits protein synthesis by combining with the smaller sub-unit of ribosomes. Both kill bacteria rather than just stopping them from reproducing.

When an antibiotic is taken the susceptible bacteria are killed. This leaves any bacteria that are resistant. These survive and do not have any competitors so make use of the available nutrients in the body and reproduce so passing on the gene for antibiotic resistance. Under normal circumstances when the antibiotic is not present, any bacteria with the gene for antibiotic resistance do not compete so well because producing a protein that is not required puts them at a disadvantage. It takes energy to replicate plasmids and to copy genes that are not essential if the antibiotic is not in the environment.

Vertical transmission is the passing on of antibiotic resistance from parent bacterium to daughter bacteria during asexual reproduction by binary fission. As the genes for antibiotic resistance are on plasmids, these can be passed from one species to another by **horizontal transmission**. Bacteria often scavenge DNA from their surroundings and this may include plasmids from dead bacteria. This makes the spread of antibiotic resistance of great concern.

☑ Study focus

In both these cases of natural selection in action the mutations do not appear in response to the selective agents. The mutations occur spontaneously and the mutant alleles may be selected if they provide some feature which is an advantage at *a certain time.*

Did you know?

Resistance to streptomycin is the result of a change in ribosome proteins so streptomycin cannot bind. Penicillin-resistance is the result of enzymes that break down penicillin.

Summary questions

1   Body colour in *B. betularia* is controlled by a gene. The allele, **B** codes for the melanic form, **b** for the speckled form. Explain how melanic forms of this species appear in populations composed entirely of the speckled forms. Use genetic diagrams to show the results of crosses between melanic forms and speckled forms.

2   Explain in detail why the frequency of melanic moths has decreased in the British Isles since the mid-1900s.

3   Explain how antibiotics act as agents of natural selection.

4   Explain how antibiotic resistance can be transmitted vertically and horizontally.

Learning outcomes

On completion of this section, you should be able to:

- describe why heritable variation is important to selection
- state that natural selection acts to maintain constancy in a population and explain how this happens
- discuss how natural selection acts as an agent of change in a population.

☑ *Study focus*

These aspects of the environment, such as predation, disease and competition, are often called *selection pressures*.

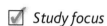 *Link*

The genetic variation is the result of mutation, meiosis and fertilisation, for further information see pages 128, 84 and 98.

Did you know?

Most crabs are crustaceans. In spite of its common name *L. polyphemus* is not related to crustaceans. It is more closely related to spiders and scorpions.

Aspects of the environment act as agents of selection. Species overproduce so there are more young individuals in the population than the habitat can support. There is competition between individuals and many starve, are eaten by predators or die of disease. These factors control the population size. The individuals that survive the early period of life when organisms are at their most vulnerable compete for nesting sites and for mates. If they are successful they get to pass on their alleles to the next generation. Alleles that are advantageous increase in frequency in the population. In a sexually reproducing species, new combinations of alleles of different genes are produced in each generation so producing as much variation as in the previous generation.

Stabilising selection

Selection does not always lead to change as it has with peppered moths; more often it maintains populations over time so that they do not change.

The horseshoe crab, *Limulus polyphemus*, is found in the seas from Nova Scotia in Canada to the Yucatán in Mexico. We know from fossils that this species has remained largely unchanged for 250 million years. **Stabilising selection** has occurred as the environment has not changed.

Research in Sweden on populations of reed warblers, *Acrocephalus scirpaceus*, provides evidence for stabilising selection. The wings of young reed warblers, reach their maximum length a few days after leaving the nest. At this age the wing length in millimetres of each bird was recorded. Each bird was identified by putting a small ring around one of its legs. When the birds were caught in net traps as adults the information on the rings was used to identify specific birds and their age. The length of time between ringing and trapping was recorded for each bird that was identified before it was released.

The mean age at trapping was calculated for birds of each wing length as shown in the table.

Wing length at ringing/mm	Number of birds trapped	Mean age at trapping/ days
less than 63	24	253
64	72	256
65	130	297
66	183	346
67	167	349
68	106	270
69	66	237
more than 70	23	199
	total = 771	

The data shows that birds with wings 66–67 mm had a greater chance of survival than those with wing lengths at the extremes of the range. Records confirm that the most common length of wing has been 66–67 mm for many generations of these birds. Scientists have discovered that genes are responsible for wing length in reed warblers so this is another example of stabilising selection.

Directional selection

Peter and Rosemary Grant have studied a population of the medium ground finch, *Geospiza fortis*, on the island of Daphne Major in the Galápagos. In the mid-1970s, the island suffered a drought and many of the plants that provide seeds for this bird died; the bird population crashed. The Grants collected data on the birds that died and those that survived.

	Birds that died	Birds that survived
mean beak length/mm	10.68	11.07
mean beak depth/mm	9.42	9.66

Only birds with beaks larger than 10.5 mm were able to eat the seeds. As the population of *G. fortis* recovered after the rains returned, the average body size and beak depth of their offspring was greater than before (an increase of 4–5% for beak depth). The bell-shaped curve had shifted to the right as **directional selection** had occurred. In 2003 to 2005 a second drought, combined with the presence of large ground finches, *G. magnirostris*, with larger beaks, drastically decreased the number of large seeds available. Now ability to eat small seeds became important. The death rate for large-beaked birds was high, leaving very few individuals with large beaks to reproduce. The beak of the medium ground finch was quickly reduced to pre-1977 size as you can see in the next table.

	Population in 2002	Population in 2005
mean beak length/mm	11.2	10.6
mean beak depth/mm	9.4	8.6

The Grants discovered that the size of the beaks is related to the genes for growth of the beak so this is an inheritable feature.

Disruptive selection

In this case a species may be spread across a geographical range and the extremes are selected in different places but the types in the middle are selected against. This gives rise to two populations which may change so much that they are not able to interbreed.

The African seedcracker, *Pyrenestes ostrinus*, is a bird that lives in equatorial forests and feeds on the seeds of sedge plants. Within the species there are:

- small-beaked birds, which have beaks 12 mm wide and feed on soft seeds
- large-beaked birds, which have beaks 15–20 mm wide and feed on hard seeds
- mega-beaked birds, which have even larger beaks so can feed on very hard seeds.

In the wet season all seeds are abundant and all bird forms feed on all types of seed. In the dry season when food becomes scarce the three forms specialise on the seeds to which they are best adapted so dividing food resources between them. Large-beaked birds specialise on a hard-seeded sedge species *Scleria verrucosa*, whereas small-beaked birds feed on the soft-seeded sedge *S. goossensii* and broaden their diet to include other soft seeds. The mega-beaked birds feed on the even harder seeds of *S. racemosa*. Birds with beaks intermediate in depth do not survive. This could be an early stage in **disruptive selection** to give two or more different populations that may become isolated to form new species.

Figure 5.7.1 A large ground finch, *Geospiza magnirostris, feeding on seeds*

Summary questions

1 Write definitions of the different types of selection described in this section: *stabilising*, *directional* and *disruptive*. Draw 'before and after' graphs to illustrate what happens to a feature that shows continuous variation for each type of selection.

2 Explain how each new generation shows more variation than its parents, yet the pattern of variation within the adult population remains the same from generation to generation in many species.

3 Predict the effects that extinction of the sedge, *S. verrucosa*, might have on populations of the African seedcracker, *P. ostrinus*, in dry seasons.

5.8 Species and how they evolve

Learning outcomes

On completion of this section, you should be able to:

- define the term *species*
- discuss the advantages and disadvantages of the biological species concept
- list the isolating mechanisms that keep species separate from each other
- explain how allopatric and sympatric speciation occur.

Did you know?

The Swedish biologist, Carl Linnaeus (1707–78) developed the binomial system for naming organisms. He classified organisms into a hierarchical system. Each species has a generic and specific name as used throughout this book.

Did you know?

The biological species concept was proposed by the naturalist and scientist, Ernst Mayr (1904–2005) in 1942.

☑ Study focus

These populations of shrimp are isolated by their behaviour.

Biological species

We have used the term species without really defining it. Linnaeus used the term to mean a group of organisms sharing common features that remained unchanged over time. By contrast Darwin thought that species could not be like this and that they changed in response to changes in their environment.

Mayr's **biological species** concept states that all individuals of a species are able to breed together to produce fertile offspring and are reproductively isolated from other species. To see if two or more individuals are members of the same species, a biologist must see if they will mate together and give fertile offspring. This is not always possible because:

- specimens collected may be dead and some are known from single specimens
- it is very difficult to observe mating behaviour in the wild, let alone check the offspring to see if they are fertile
- fossil species cannot breed
- some species reproduce asexually or by parthenogenesis (unfertilised egg cells give rise to the next generation).

As a result of these difficulties most scientists use the **morphological species** concept. They define species by the possession of common distinctive features of morphology (outward appearance), behaviour, physiology, biochemistry and anatomy.

Isolation of different species

There are various methods that prevent individuals from different species breeding with each other. These **isolation mechanisms** are summarised in the table opposite.

Speciation

New species arise from existing species. The process by which this happens is known as **speciation**.

Allopatric speciation occurs when populations of a species are physically separated and are in two different geographical areas. **Sympatric speciation** occurs when the two or more populations are in the same geographical area.

Allopatric speciation

If individuals of a species migrate to occupy a new area, they are exposed to different selection pressures compared with the area that they have left. Over time this may lead to changes in the isolated population to the extent that they cannot interbreed with the original population. Populations of the snapping shrimp, *Alpheus*, were separated when the isthmus of Panama was formed 3 million years ago. Populations on either side of the isthmus are very similar, but when males and females from these two populations are put together they 'snap' at one another and do not mate.

Sympatric speciation

Speciation may occur within a population. Usually this is an abrupt change in a species, so that individuals are not able to interbreed. Hybridisation sometimes occurs between different species of plant.

Isolating mechanism		Example
Pre-mating isolation methods		
geographical/ecological	features such as rivers, lakes, mountains, lowlands, forests separate populations so they never or rarely meet	the three sub-species of elephants in different parts of Africa are separated geographically
temporal or seasonal	breeding occurs at different times of the year	two species of pine in California release pollen in February and April so do not pollinate each other
reproductive behaviour	courtship rituals differ between species so males and females do not respond to signals and mate (assortative mating)	female anolis lizards recognise the displays of males of the same species and ignore males that show different displays of dewlap and head bobbing
mechanical	structural differences mean that sex organs of males and females of different species are incompatible	exoskeleton of arthropods gives a 'lock and key' arrangement of male and female genitalia – only those of the same species 'fit together'
post-mating isolation methods		
production of a hybrid prevented	fertilisation does not occur	sperm of *Drosophila viridis* is immobilised in tract of female *D. americana*
		pollen fails to grow pollen tubes on stigma of different species
	embryo fails to develop	crosses between goats and sheep
breeding of hybrids prevented	offspring are sterile (often because homologous chromosomes cannot pair in meiosis)	cross between horse (female) and donkey (male) gives sterile mule
	no viable individuals are produced	when *Drosophila melanogaster* mates with *D. simulans*, no viable males are produced

The hybrids are sterile and not able to breed with either parent. However, if the chromosome number doubles by a failure of meiosis they become fertile. An example of this speciation is polyploidy in cord grass, a plant that grows in saltmarshes. In the UK in the early 19th century there was only one species, *Spartina maritima* ($2n = 60$). The North American species, *S. alterniflora* ($2n = 62$), was introduced accidently in the middle of that century and hybridised with the native species to produce the sterile *S. townsendii* ($2n = 62$) which spread vegetatively. Polyploidy occurred to double its chromosome number to become the fertile species, *S. anglica* ($2n = 124$).

For sympatric speciation to occur hybrids must be fertile and able to breed with other such hybrids, but not breed with either parental species. A hybrid must also be adapted to survive in its habitat or have a habitat that is not shared with either parental species. An example is the Lonicera fly which is a hybrid between the species *Rhagoletis mendax*, which has a maggot that feeds on blueberries, and *R. zephyria*, with a maggot that feeds on snowberries. The hybrid, known as *R mendax* × *zephyria* has maggots which feed on the fruits of honeysuckle, *Lonicera*, and so does not compete with either parent species.

✅ *Study focus*

Darwin thought that speciation is a long, gradual process. *S. anglica* is an example of a species that evolved very quickly over just a few years.

Summary questions

1 Define the term *biological species*.

2 Explain why: **i** it is often difficult to apply the biological species concept; **ii** the morphological species concept is easier to apply in practice.

3 Distinguish between allopatric and sympatric speciation.

Learning outcomes

On completion of this section, you should be able to:

- analyse data about natural selection

- apply knowledge and understanding to explain results of investigations into selection

- suggest how variation in populations might be investigated to discover the agents of selection.

Many scientists have investigated examples of selection. The results of some of these investigations are presented here.

Kettlewell's experiments on peppered moths

Bernard Kettlewell investigated the effect of predation by birds as the selective agent in maintaining populations of peppered moths in the 1950s and 1970s. He trapped and reared melanic and non-melanic forms, marked them and then released them in a wood in an industrial area of Birmingham and a wood in a rural area where the trees were covered in lichen. After a few days he used a moth trap to recapture the moths. His results are in the table.

1 Describe and explain the data shown in the table.

Kettlewell observed the behaviour of predatory birds and filmed them feeding on the peppered moths that were least well camouflaged against the bark of the trees. His observations provided evidence that visual predation by birds was the agent of selection in both areas.

Site of woodland		Number of peppered moths		
		non-melanics	melanics	total
non-polluted, rural area	marked and released	496	473	969
	recaptured	62	30	92
	% of marked moths recaptured	12.5	6.3	
polluted, industrial area	marked and released	64	154	218
	recaptured	16	82	98
	% of marked moths recaptured	25.0	52.3	

Melanism is not unusual in animal species. An example is the variation within populations of the bananaquit, *Coereba flaviola*, on Grenada and St Vincent. The more common form is mainly yellow with some black marking. Melanic forms are completely black and on Grenada and St Vincent they are only found in moist forest at low and high altitudes whereas the yellow forms live on disturbed dry lowland habitat. Melanics have a stronger preference for shade than the yellow birds. There is no evidence for assortative mating in which each form mates with birds of the same colour.

2 Suggest how you might investigate melanism in bannaquits to discover how it is inherited and how it is maintained in populations on Grenada and St Vincent.

Banded snails

Banded snails, *Cepaea nemoralis* and *C. hortensis*, live on the ground amongst leaf litter and ground vegetation. They exist in three different colours: brown, pink and yellow. Some snails have bands, others do not. These two features are controlled by alleles at two separate loci on different chromosomes. In a survey, *C. nemoralis* snails were collected from woodland where the leaf litter was dark brown and adjacent grassland where the background colour is more variable, but mostly yellow and green. The table shows the percentage of yellow-shelled snails and unbanded snails in three samples from each area.

Habitat	Sample	Percentage of yellow snails in sample	Percentage of unbanded snails in sample
woodland	1	12	88
	2	21	77
	3	12	70
grassland	1	79	21
	2	58	14
	3	83	22

These results are very similar to those obtained by surveys in similar habitats over many years.

Birds known as thrushes are common predators of *C. nemoralis*. The birds hold them in their beaks and strike them against a stone. Broken shells are found around the stones. Equal numbers of banded yellow and unbanded brown snails were collected and marked before being released into the centre of each habitat. A count was made after a few days on broken snail shells around the stones in each habitat. The table shows the results.

Habitat	Banded yellow shells	Unbanded dark shells
grassland	23	76
woodland	68	30

3 Use the information to explain how selection pressures act to maintain different types of *C. nemoralis* in the two habitats.

Predation of 'spaghetti worms' by birds

You can test the effect of visual predation by putting out food for birds. Scientists have developed a method to test the theory that visual predation by birds could be an agent of selection by making different coloured food with pastry and food colouring. Sheets of aluminium were painted grey, green and brown and were placed on the ground. 20 brown and 20 green pieces of 'prey' were placed on each tray and left until about half the pieces had been eaten by the birds. The numbers eaten were recorded. The results are in the table.

Guppies in Trinidad

Guppies, *Poecilia reticulata*, are small fish that live throughout the Caribbean. David Reznick has studied the effects of predation on populations of these fish in the Aripo river system in Trinidad.

The guppies living in rivers with predators grow up faster and mature at smaller sizes; they also reproduce at a younger age than those in rivers without predators. Guppies in rivers with predators breed as soon as possible because they are at risk of being eaten. The investigations involved moving individuals from each population and putting them into rivers which did not have guppies. Some of these new rivers had predators in them, others did not. The experiments were designed so that populations were moved from rivers with predators to rivers without predators, and from rivers without predators to those with predators. As a control, populations were moved from rivers with predators to different rivers with predators, and the same was done for populations from rivers lacking predators.

The fish were monitored for eleven years. Over that time in rivers without predators the body size of the guppy population had increased and sexual maturation occurred later; in rivers with predators the fish were smaller and matured earlier. This showed that the guppy populations had changed to maximise their chances of reproduction by adopting different strategies depending on the presence or absence of predators.

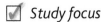

Study focus

Look online for photos of banded snails *Cepaea nemoralis* and *C. hortensis* in their natural habitat.

Figure 5.9.1 *These snails with different banding patterns are all members of the same species,* Cepaea hortensis

Background colour	Number of 'prey' eaten		Percentage of 'prey' eaten	
	green	brown	green	brown
grey	147	118	55.5	44.5
green	56	72	43.8	56.2
brown	62	31	66.7	33.3

4 Describe and explain the data on predation of 'spaghetti worms' shown in the table.

5 Explain how you would continue the experiment to find out the effect of selection on a population of 'spaghetti worms' over time.

6 These are all examples of selection, not speciation. Explain:

a how speciation can occur over a few generations

b how you can show that speciation has occurred.

This section contains SAQs on variation and natural selection. You can find MCQs on the CD that accompanies this book.

1 a Describe the differences between continuous variation and discontinuous variation as seen in the phenotype of a NAMED organism.

   b Explain how the genetic control of discontinuous variation differs from that of continuous variation.

   c Outline the importance of inheritable variation for natural selection.

2 a Explain the advantages to modern agriculture of having crop varieties which are genetically uniform.

   b Outline the disadvantages of growing genetically uniform crop varieties over large areas of land.

   c Suggest reasons for keeping many different breeds of livestock, such as Jamaica Hope cattle and Barbados blackbelly sheep, for the future of farming.

3 a Describe TWO examples of the effects of the environment on the phenotype of an organism.

   b Explain how i the interaction of alleles at one locus, and ii the interaction of alleles at different loci can influence the phenotype of an organism.

   c Explain the importance of phenotypic variation for natural selection.

4 a Define the term *mutation*.

   b With reference to sickle cell anaemia and Down syndrome, explain the difference between gene and chromosome mutation.

   c Explain why sickle cell anaemia is a serious health problem in parts of Africa and among people elsewhere in the world that are descended from African populations.

5 The black-faced grassquit, *Tiaris bicolor*, is a small bird found throughout the Caribbean and the coasts of Venezuela and Colombia. It is related to Darwin's finches of the Galápagos Islands.

   a Describe how variation in wing length in a population of *T. bicolor* on a Caribbean island would be investigated, recorded and presented.

   b Explain how it is possible to determine if the population on that island belongs to the same species as populations on the South American mainland.

   c It is thought that a small population of birds such as *T. bicolor* colonised the Galápagos Islands. Explain how such a small population may have given rise to the 14 species of finch now found on those islands.

6 *To answer this question you will need to use the DNA genetic dictionary on page 67.*

   a The following is a sequence of DNA triplets on the template strand of the gene for an octapeptide:

      CTA ATG TAC CCA ACC TAC CTA AAG

      State what happens to the amino acid sequence of the peptide produced in translation if the following gene mutations occur to the SECOND DNA triplet:

      i   a deletion of the first base
      ii   the first base changes from A to T
      iii the triplet is deleted
      iv   the addition of the base pair T at the beginning of the triplet
      v   the third base changes from G to C.

   Mutations that do not change the amino acid coded by a triplet are called neutral mutations. Other types of mutation are substitution and frameshift mutations.

   b Classify the gene mutations you have described in a as substitution, frameshift or neutral.

   c Explain how the behaviour of chromosomes during meiosis is a cause of Down's syndrome.

7 a Define the term *biological species*.

   Seahorses are small marine fish. Reproduction in seahorses is unusual as it is the male that becomes pregnant and has a brood pouch on its tail. Larger males have larger brood pouches. A female seahorse transfers her unfertilised eggs into the male's brood pouch. He releases sperm onto the eggs and they are fertilised. He keeps the developing young for a period of several weeks and then gives 'birth'.

   A male and a female seahorse remain together as a pair for the mating season. Within a population females select males by size. Large females pass their eggs to the brood pouches of large males and small females pass their eggs to small males. The selection of mates in this way is known as *assortative mating*. Very few seahorses of intermediate size survive and breed.

Two closely related species of seahorse, *Hippocampus erectus* and *H. zosterae* are found in the same coastal waters of Florida. *H. erectus* has a mean length of 120 mm, *H. zosterae* has a mean length of 20 mm.

b Name the type of speciation that occurs when two species exist in the same geographical area.

c Use the information provided to explain how *H. erectus* and *H. zosterae* may have evolved from another species of *Hippocampus* in the coastal waters of Florida.

d Explain how species remain reproductively isolated other than by assortative mating.

8 Gonorrhoea is a bacterial disease. Antibiotic sensitivity tests are carried out on samples of bacteria taken from people with gonorrhoea to ensure that they are treated with an appropriate antibiotic.

a State how bacteria become resistant to antibiotics.

b Explain how antibiotic resistance is transmitted from one bacterial cell to others.

c Explain why it is important to carry out antibiotic sensitivity tests before prescribing a course of antibiotics.

d Antibiotic resistance is a serious medical problem worldwide. Suggest ways in which this problem may be overcome.

9 a Define the term *speciation*.

The plant *Euphorbia tithymaloides* is found throughout Central America and the Caribbean. It is thought that the plant spread from Central America in two directions – along the north via Cuba and Jamaica and from the east via Trinidad. Populations on adjacent islands interbreed successfully, but in St. Croix in the north-east there are two populations that do not interbreed.

b Explain the significance of the following statements.

i Populations on adjacent islands interbreed successfully.

ii There are two populations on St Croix that do not interbreed.

c Explain why heritable variation is important for natural selection.

10 Many different species of cichlid fish live in Lake Victoria in Africa. Some species live in shallow water and others live in deeper water. The wavelengths of light that penetrate water influence the features of fish that live in lakes. Blue light does not penetrate far into lake water; red light penetrates much further.

The table summarises the features of males and females of these species.

Habitat	Body colour of males	Retina in eyes of females
shallow water	blue	detects blue light
deep water	red	detects red light

Body colour and colour vision are both inherited features. Females select the males that they mate with and prefer coloured males.

a i The ancestors of red and blue cichlid fish were brown.
State how the different body colours of the males arose.

ii Suggest the advantages of different cichlid fish being able to detect blue and red light.

b Many species of cichlid fish have evolved in lakes in Africa. Use the information provided to explain how speciation may have occurred in these lakes.

c Lake Victoria receives considerable pollution from the surrounding area, which makes the water cloudier and reduces the penetration of blue light. Suggest and explain the likely long-term effects of the cloudy water on the species of red and blue cichlid fish.

11 a Explain why body mass is an example of continuous variation.

Three separate populations of mice on different islands across the world were studied over a long period of time. The researchers collected data on body mass. At the beginning body mass of all the mice showed a normal distribution and the mean body mass of each population was about the same. At the end of the study, these results were obtained:

Population A no change to the variation in body mass

Population B the mean mass increased, but with fewer small mice and more larger mice

Population C there were many small mice and many much larger mice, but none with the mean mass of the original population.

b Draw graphs for populations A, B and C to show the variation in body mass at the beginning and at the end of the study.

c Name the type of selection that occurred to each population.

3 Reproductive biology

1.1 Asexual reproduction

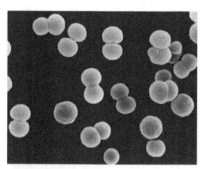

Figure 1.1.1 *These bacteria are dividing by binary fission*

∞ Link

Remember that prokaryotes do not have a nucleus or linear chromosomes like those of eukaryotes, see page 32.

Did you know?

Multiple fission occurs in some organisms; while in red blood cells the malarial parasite divides into many tiny infective stages that are picked up by a female *Anopheles* mosquito when it feeds on blood.

☑ Study focus

Remember that yeast does not divide by *binary fission* as the bud is much smaller than the parent cell.

Asexual reproduction is the production of new individuals from a single organism without the fusion of gametes. All the individuals formed in this way form a **clone** of genetically identical organisms. There may be some limited variation amongst the organisms in a clone that is caused by random mutation. In prokaryotes and eukaryotes replication of DNA occurs before cells divide and replication is usually free from errors. In eukaryotes, mitosis occurs so that daughter cells receive the same number and types of chromosomes. This maintains genetic stability.

Examples of asexual reproduction are:

- binary fission in bacteria
- budding in yeast and *Aiptasia* (a sea anemone)
- fragmentation in *Spirogyra* (a filamentous alga)
- spore production in *Rhizopus nigricans* (a mould fungus).

Asexual reproduction in prokaryotes

Binary fission in bacteria

Asexual reproduction in prokaryotes, such as bacteria, is simpler than in eukaryotes but shares some similarities. This type of reproduction is known as **binary fission** as each cell splits into two. Bacterial cells absorb food and grow larger. When a cell reaches a certain size it starts to divide:

- the circular chromosome replicates
- the two chromosomes separate while held on to the cell surface membrane
- two new cell membranes form across the middle of the cell
- cell wall material is formed between the new membranes
- the cells remain attached for a while and then split apart.

Some bacteria, such as *Escherichia coli*, divide every 20 minutes under optimum conditions.

Asexual reproduction in eukaryotes

Budding in yeast

Yeast cells grow in size but do not divide equally in half. At a certain size:

- the 16 linear chromosomes replicate
- the nucleus begins to divide by mitosis, but the nuclear membrane remains intact and does not break up as shown in the diagram on page 79
- a small swelling appears at the side of the yeast cell to form a bud
- one of the daughter nuclei enters the bud
- the bud remains attached for a while and then breaks off leaving a bud scar on the parent yeast cell.

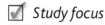

Budding in *Aiptasia*

Aiptasia is a sea anemone, which reproduces asexually by a form of budding called pedal laceration. Small groups of cells from the base of the animal grow into small buds. After about a week these separate from the animal and each then develops a mouth surrounded by tentacles so it can feed itself.

Fragmentation in *Spirogyra*

Spirogyra is one of many types of multicellular, filamentous alga that grows on the surfaces of ponds. A short single chain of cells grows in length. Any cell in the chain can enlarge, divide by mitosis to form two cells and make the chain longer. Growth does not happen just at the ends of the filaments. New filaments form when lines of weakness develop between cells. Any disturbance to the filaments causes them to break. When conditions are perfect for growth, algae like *Spirogyra* can cover the surfaces of freshwaters in a thick growth called blanket weed. Many flowering plants also reproduce by fragmentation, for example the Mexican Hat plant, *Kalanchoe daigremontiana* grows tiny plantlets around its leaves. After a while these break off and fall to the ground.

Spore formation in *Rhizopus*

Leave some damp bread exposed to the air for 30 minutes, place in a Petri dish and look at it every day for the next few days. You will see a growth of mould over the surface of the bread. This is bread mould, a type of fungus. The common bread mould, *Rhizopus nigricans*, has branching filaments, or hyphae, which spread over the surface and grow into the bread. If you look at the growth with a hand lens you can see lots of tiny black pin heads. These are the structures carrying out asexual reproduction for the fungal body, the **mycelium**, which you can see as a mass of white filaments.

Fungi exhaust their food supplies, often very quickly, so they need to produce spores to colonise new sources as follows:

- hyphae known as **sporangiophores** grow up into the air
- the tip of each sporangiophore swells up to form a **sporangium**
- haploid nuclei move from the hyphae into the sporangium and divide by mitosis
- the columella – an extension of the sporangiophore – pushes up into the sporangium separating it from the rest of the mycelium
- cytoplasm forms around a group of several nuclei to form a spore
- a wall forms around each spore
- when the spores are ready, the sporangium splits open and the spores dry and are blown away in air currents.

Fungal spores may be carried thousands of miles as they are so small and light; most will not land on a food source, so there is great wastage.

Many organisms, both prokaryote and eukaryote, produce spores. They vary considerably in their structure so the term *spore* is difficult to define. They are best thought of as microscopic, reproductive bodies that contain cytoplasm and one or more nuclei. They are often dispersed, but not always as some are for surviving harsh conditions and not really for dispersal in the same way as the spores of *R. nigricans*. The embryo sacs of flowering plants are spores although they are not released nor are they for surviving hash conditions (see page 151).

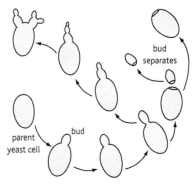

Figure 1.1.2 *These yeast cells are budding. The nucleus divides and one of the daughter nuclei enters the bud before it breaks off.*

Did you know?

Rhizopus nigricans and *R. tritici* cause soft rots in sweet potatoes. These fungi damage the potatoes so they cannot be sold.

Summary questions

1 Define the following terms: *asexual reproduction, binary fission, budding, fragmentation* and *spore.*

2 According to the endosymbiosis theory mitochondria are derived from prokaryotes. The number of mitochondria increases during interphase of the cell cycle. Describe how this happens.

3 Describe the functions of the following in asexual reproduction of *R. nigricans*: sporangiophore, mitosis, sporangium and spore.

4 Suggest how you might prevent the contamination of stored sweet potato tubers by *R. nigricans.*

5 Discuss the advantages and disadvantages of asexual reproduction for microorganisms, such as bacteria, fungi and algae.

6 The sea anemone, *Aiptasia* is occasionally introduced by accident into marine aquaria. It can very quickly spread throughout aquaria causing harm to other animals. Suggest why it spreads so easily.

⚮ Link

Meristematic tissue consists of undifferentiated cells that divide by mitosis. They are equivalent to animal stem cells as referred to on page 76.

Did you know?

The Gros Michel variety of banana was attacked by a fungal disease known as Panama disease in the 1950s. Another variety, Cavendish, is grown in plantations all over the tropics. A new strain of Panama disease (tropical race four) that infects the Cavendish variety has emerged in South-East Asia and may well spread to the Caribbean and Central America.

Many flowering plants reproduce asexually. This is known as **vegetative reproduction** or **vegetative propagation** as it involves producing new individual plants by modification of vegetative growth rather than growth for reproduction involving flowers. Propagation is a word meaning multiplication or spreading. Plants can produce new individuals by growth from:

- leaves, e.g. *Kalanchoe daigremontianum* and African violet, *Saintpaulia ionantha*
- stems, e.g. Irish potato, *Solanum tuberosum*, and ginger, *Zingiber officinale*
- roots, e.g. breadfruit, *Artocarpus altilis*, and dandelion, *Taraxacum officinale*.

Ginger plants have a swollen horizontal stem known as a **rhizome**. The stem has short **adventitious roots** that grow from it as it does not have a main root. Leaves grow from leaf buds on the rhizome. There are **axillary buds** at the point where the leaves grow. These buds have meristematic tissue, which grows to provide branches on the rhizome. The rhizome system can spread through the soil and cover much of the ground. Eventually the rhizome will fragment into several separate plants.

Structures such as rhizomes are swollen with stored energy reserves and nutrients for survival over unfavourable times of the year when it is too hot, too dry or too cold for growth. Most structures that reproduce vegetatively are like this. Structures, such as rhizomes, tubers and bulbs, store materials for **perennation**.

Plant hormones

The growth and development of plants are influenced by growth substances that are often called plant hormones. There are three main types of hormone.

Type of plant hormone	Example	Source within the plant	Effects on growth
auxin	indolyl acetic acid (IAA)	shoot apex	stimulates elongation of cells by 'softening' cell walls so they are stretched when cell absorbs water
cytokinin	kinetin	root apex	stimulates cell division
gibberellin	gibberellic acid (GA$_3$)	roots and young leaves	stimulates cell elongation and growth in length of stem

Synthetic plant hormones are used to control the growth of crop plants:

- the auxin naphthaleneacetic acid (NAA) is used as a rooting compound for cuttings
- gibberellins are sprayed on seedless grapes in California to increase the size of each grape and increase the distance between grapes so reducing chance of disease spreading through the crop
- kinetin is used in tissue culture to stimulate cell division.

Artificial propagation

Cuttings are taken from plants by removing stems, root or leaves from a parent plant that has good features that are worth propagating. This can be done by cutting:

- a leaf at the point where it meets the stem, e.g. African violet
- across a side shoot just below the point where a leaf joins the stem, e.g. Joseph's coat, *Coleus*, and sugar cane, *Saccharum officinarum*
- across the root, e.g. breadfruit, *Artocarrpus altilis*.

Stem cuttings may be dipped into the NAA and then placed into a suitable soil or compost for growing cuttings. After a while adventitious roots grow from the base of the cutting and the new plant can support itself.

Not all cuttings are genetically identical to the parent plant from which they have been taken. Some varieties of cultivated plants are chimeras, which means that they are composed of cells of two different genotypes. Root cuttings of thornless blackberry plants grow with thorns as the cells from which the shoots grow have the genotype for the thorned condition.

Plants that naturally produce structures for vegetative propagation can be divided. Banana trees have suckers at their base which are removed for propagation. This is the only way cultivated bananas, which are triploid and therefore sterile, can be propagated.

There are advantages of vegetative reproduction:

- new plants are easy to establish (many seeds are not viable and do not grow)
- most plants are clones and genetically uniform
- plants have uniform appearance which gives a reliable supply
- plants are uniform size which helps harvesting and packing

The main disadvantage is that genetically uniform crops are at risk of the same pests and diseases for which they have no resistance. If there is an epidemic of a pest or disease there is a chance that all plants with the identical genotype will be wiped out.

Micropropagation

Many commercially important plants are propagated using methods of **tissue culture**. Small pieces of tissue are removed and cultured in a liquid medium or on a solid medium. The media contain sucrose as a source of energy, nutrients so the plant cells can make all the biological molecules they need and plant hormones to regulate the type of growth. At the beginning of culturing the number of cells increases, but at the end cells are stimulated to differentiate into stems, roots and leaves.

Aseptic technique is used to ensure that no contaminants ruin the stock of plants. This involves sterilising the media, using sterile containers and implements; spores are removed by filtering the air and staff can take precautions such as wearing appropriate clothing, masks and gloves.

Summary questions

1   Suggest the reasons why some plants are propagated vegetatively rather than by seed.

2   Explain why plant hormones are used in tissue culture.

∞ Link

Auxins stimulate root growth and cytokinins stimulate shoot growth, refer to Question 7 on page 157.

Did you know?

Scientists once thought that during differentiation cells only kept the genes they needed to carry out the functions of specialised cells and that they lost some or all of the rest. Growing whole plants from differentiated cells shows this is not the case.

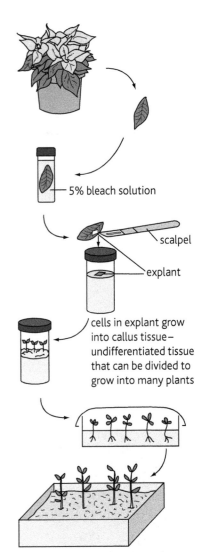

5% bleach solution

scalpel

explant

cells in explant grow into callus tissue – undifferentiated tissue that can be divided to grow into many plants

Figure 1.2.1 *Tissue culture is used to produce large numbers of plants with commercially important features*

Learning outcomes

On completion of this section, you should be able to:

- explain the term *sexual reproduction*
- describe the structure of a typical insect pollinated flower and the detailed structure of an anther and an ovule
- describe the formation of pollen grains and embryo sacs.

∞ Link

Remind yourself of the life cycle of flowering plants on page 81.

☑ Study focus

Look for photos and drawings of insect pollinated flowers so that you can identify the different parts. Then find some insect pollinated flowers and inspect them carefully with a hand lens.

Sexual reproduction involves the fusion of gametes. In flowering plants meiosis occurs in the life cycle in the production of spores *before* the fusion of gametes. The spores are more complex than those produced by *Rhizopus* (see page 145).

Most flowers are **hermaphrodite** as they have both male and female structures. The **stamen** is the male part of the flower and the **carpel** is the female part.

Each stamen is composed of a filament and an **anther**. Filaments contain xylem and phloem to transport water, ions, sucrose and amino acids to the anthers. Each anther is composed of four **pollen sacs** where pollen grains develop. The diploid cells, which divide by meiosis to form pollen grains, are **pollen mother cells**. If you look at sections of anthers at different stages of development you can see cells in the stages of meiosis. At the end of meiosis the four cells form tetrads each of which develops into a mature pollen grain. Follow the production of pollen grains in Figure 1.3.1 and make sure you look at prepared slides or photomicrographs of pollen sacs.

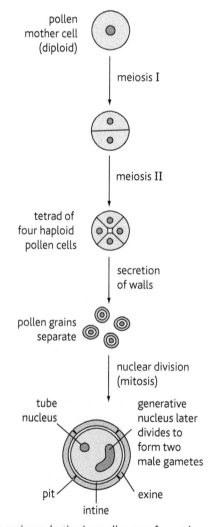

Figure 1.3.1 *Pollen grain production in a pollen sac of an anther*

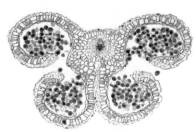

Figure 1.3.2 *Cross section of an anther showing release of mature pollen grains from the four pollen sacs*

Pollen grains have a thin inner wall, the **intine**, made of cellulose and a thick outer wall, the **exine**, made of sporopollenin which is tough and waterproof. There are pits in the exine. The haploid nucleus within the pollen grain divides by mitosis to form a **generative nucleus** and a **tube nucleus**.

At the base of the carpel is the ovary, which is swollen to enclose one or more ovules. Each ovule is attached to the inside of the ovary so that it receives water and nutrients. The integuments enclose the nucellus and a diploid **embryo sac mother cell** that divides by meiosis to form four haploid nuclei. Three of these degenerate leaving one nucleus to divide by mitosis to form eight haploid nuclei. These move to occupy the positions inside the embryo sac that you can see in Figure 1.3.3. The two polar nuclei may fuse to form a diploid nucleus.

Did you know?

Pollen grains are very resistant. If you take soil, lake or peat samples from different depths it is possible to tell how the vegetation has changed in an area over time. This is useful in archaeology.

☑ Study focus

Degenerate here means that the nuclei do not divide again, do not function and probably break down. Degenerate is also used to describe the genetic code where it means that there are more codes than needed for 20 amino acids (see page 67).

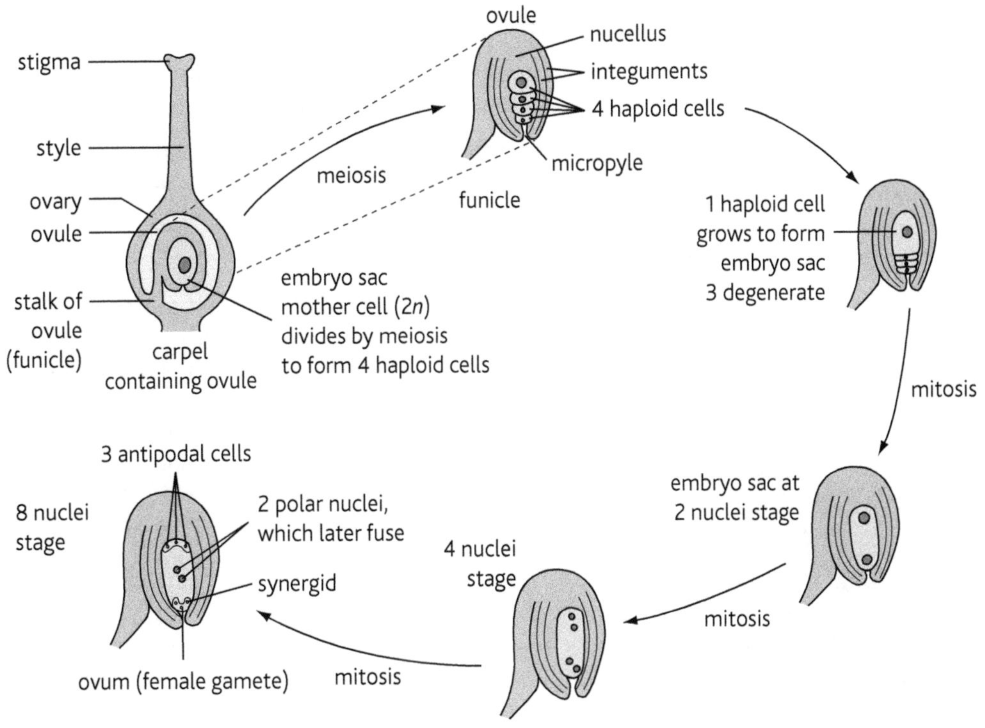

Figure 1.3.3 *The development of the embryo sac in the carpel*

Summary questions

1   Describe the events that occur to produce a mature pollen grain and a mature embryo sac.

2   Explain the advantages of meiosis occurring in the flowering plant life cycle.

3   Suggest why three nuclei degenerate following the meiotic division of the embryo sac mother cell.

4   Palynology is the study of spores, including pollen grains. Suggest how it is possible to use pollen grains to show how vegetation has changed over historical time.

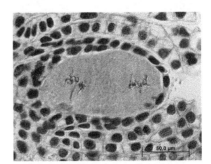

Figure 1.3.4 *This shows an embryo sac mother cell at the metaphase II stage of meiosis*

1.4 Pollination

Learning outcomes

On completion of this section, you should be able to:

- define the term *pollination*
- distinguish between self- and cross-pollination
- explain the different ways in which flowering plants promote cross-pollination
- outline how cross-pollination occurs.

Did you know?

The world's smelliest flower is the titan arum, *Amorphophallus titanum*, which grows in Indonesia. It gives off the smell of rotting flesh to attract sweat flies as pollinators.

Figure 1.4.1 *Many orchid species deceive their pollinators by attracting them with scent but not having nectar as a reward. Others look like female insects, so males come and mate with them!*

Self- and cross-pollination

Pollen grains are released from pollen sacs when the anthers break open. The opening of the anthers is known as dehiscence. Pollen grains are the way in which the male gamete is transferred to the female gamete, which is in the embryo sac inside an ovule. Fertilisation is internal as it is in mammals (see page 164). The first stage in the transfer is **pollination**, which is the transfer of pollen grains from anther to stigma. This may be as simple as anthers opening to release pollen directly onto the stigma as happens in **self-pollination**.

Self-pollination is the transfer of pollen from the anther of a flower to a stigma in the same flower or in another flower on the same plant.

Flowers of the groundnut, *Arachis hypogoea*, do not open and pollen grains are transferred directly from anthers to stigma.

Many flowers are adapted for **cross-pollination** so that pollen is transferred between flowers on different plants in a specific way.

Cross-pollination is the transfer of pollen from the anther of a flower on one plant to a stigma of a flower on another plant of the same species.

There are at least five agents of cross-pollination as shown in the table.

Agent of pollination	Ways in which an example is adapted for pollination
wind	maize, *Zea mays*, has separate male and female flowers; male flowers (tassels) are at the top of the plant to release pollen into the air to promote dispersal
insect	periwinkle, *Catharanthus roseus*, has long narrow flowers with nectar at the base – adapted for pollination by butterflies that have long mouthparts
bird (e.g. hummingbird)	false bird of paradise flowers, *Heliconia spp.*, are red which is a colour birds can see, but insects cannot; these are long, tube-shaped flowers
water	seagrass, *Thalassia testudinum*, releases pollen into the water that is carried from male flowers to female flowers on separate plants
bat	seabeans, *Mucuna spp.*, are vines that have large pungent smelling flowers to attract bats

Insect pollinated flowers have features that attract insect pollinators:

- scent
- brightly coloured petals
- honey guides – lines on the petals directing to the centre of the flower
- nectar – solution of sugars to provide insects with a 'reward' for visiting flower
- pollen – as a source of food for some insect pollinators
- landing stages on the petals.

The structure of insect pollinated flowers is adapted for insect pollination. The stigmas are usually above the anthers so that they do not self-pollinate, the pollen grains are sticky and often spiky so they stick to an insect's body. The surfaces of the stigmas are also sticky and are covered in small projections to hold the pollen grains.

Orchids practise deception. Some orchid flowers look as if they should offer food in the form of nectar, but they do not. Others practise sex deception by mimicking the appearance of female insects and emitting an appropriate scent. Male insects are attracted to the flowers and attempt to mate with them thereby transferring pollen.

Flowering plant species have a variety of ways to promote cross-pollination and avoid self-pollination.

- **Dioecy.** Seagrape, *Coccoloba uvifera*, is an example of a **dioecious** plant that has separate sexes with trees producing either male or female flowers. Male trees need to be nearby for the female trees to produce fruits. Self-pollination is impossible.
- **Dichogamy.** Stamens and carpels do not open at the same time so reducing chances of cross-pollination. *Aristolochia* species show **protogyny** in which the stigma matures and accepts pollen before the anthers open. *Bidens pilosa* is a species that shows **protandry** in which the anthers mature and release pollen grains before the stigma is ready to accept them. In many dichogamous species there is a period of overlap when the anthers are open releasing pollen and the stigmas are ripe to allow self-pollination as a fail safe.
- **Heterostyly** is the existence of plants with two types of flower within a species. Some plants have flowers with the anthers above the stigma as the style is very short; other plants of the same species have flowers with anthers below the stigma as the style is long. This ensures that insect pollinators pick up pollen on different parts of their bodies and transfer pollen from
  - flowers with short styles to flowers with long styles
  - flowers with long styles to flowers with short styles.
- **Self-incompatibility.** Plants also have genes that determine whether pollen grains germinate and grow on stigmatic surfaces. The S gene has multiple alleles (see page 102). If a pollen grain has an allele that is the same as one on a stigma it will not germinate.
- **Male sterility.** Some mutations result in the failure to produce pollen grains. This can be the result of mutations in genes on chromosomes in the nucleus and also genes in mitochondria. Plants that have the mutant allele cannot self-pollinate so have to be cross-pollinated. Plant breeders make use of male sterility by breeding maize varieties that are hybrids between existing varieties.

Study focus

Find some flowers that are insect pollinated and find the features that are listed here. Also find some electron micrographs of stigmatic surfaces.

Study focus

Apply your knowledge and understanding of meiosis and self-incompatibility to Questions 8 to 10 on page 155.

Summary questions

1 Define the following terms: *pollination*, *dioecy*, *dichogamy* and *self-incompatibility*.

2 Distinguish between the following pairs of terms: self-pollination and cross-pollination; anther and stigma; protandry and protogyny.

3 Describe the different ways in which flowering plants promote cross-pollination.

4 Explain the advantages of cross-pollination.

5 Suggest why dioecy is rare in plants, but common in animals.

6 The stamens of some cabbage plants cannot produce functional pollen due to a mutation that affects mitochondrial activity in the anthers. Explain why mitochondria are required to produce functional pollen.

7 Explain the advantages of male sterility to plant breeders.

Learning outcomes

On completion of this section, you should be able to:

- describe the sequence of events between pollination and fertilisation in flowering plants
- explain the significance of double fertilisation
- describe the development of
  a a zygote into a mature embryo plant, **b** an ovule into a seed, and **c** an ovary into a fruit.

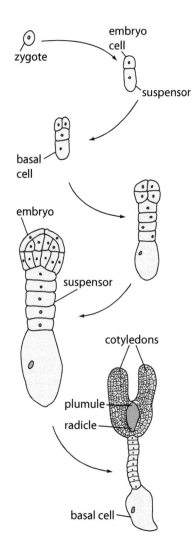

Figure 1.5.1 *The growth of the embryo of Shepherd's purse,* Capsella bursa-pastoris

After pollination

Pollen grains land on stigmas. They absorb water and sucrose and germinate. A pollen tube emerges from one of the pores through the exine and grows into the stigmatic tissue and then down through the style towards the ovule.

☑ Study focus

Both pollen grains and seeds are said to germinate. Take care when you use the word 'germination' that you are using it in its correct sense.

The pollen tube may be attracted by molecules secreted by the style and/or ovule. This is an example of **chemotropism** in which the direction of growth of the tube is influenced by chemicals.

Genes in the tube nucleus are transcribed and translated to provide proteins needed for the growth of the tube. Enzymes are secreted by the tube to digest a pathway. During the growth the generative nucleus divides by mitosis to give two haploid male gamete nuclei. The tube enters the micropyle and the tip of the tube breaks down so the male gamete nuclei can enter the embryo sac. One of the nuclei fuses with the ovum nucleus to form the diploid zygote nucleus and the other fuses with the diploid nucleus in the centre of the embryo sac to form a triploid **endosperm** nucleus. This is **double fertilisation**.

If there is more than one ovule in the ovary then they will all be fertilised by male gamete nuclei from different pollen tubes which in turn form from different pollen grains. These pollen grains could have come from different plants of the same species visited by animal pollinators or, in the case of wind pollinated flowers, blown by the wind. This gives the opportunity for much genetic variation amongst the seeds in the ovary.

Double fertilisation

The significance of double fertilisation is that the triploid endosperm nucleus divides by mitosis to give a tissue that surrounds the developing embryo. In some plants, such as the legumes, the endosperm is used up before the embryo is mature. In cereals, such as maize, wheat and rice, the endosperm remains after the embryo matures and remains in the grain. The significance for us is that these crop plants, and rye, oats, sorghum and millet, provide most of the staple foods in the human diet.

After fertilisation

Once fertilisation has occurred the zygote divides by mitosis to form an embryo with a basal cell and suspensor cells. It grows into the endosperm and obtains water, sucrose and ions from the xylem and phloem through the endosperm. This is more efficient than receiving water and nutrients from the nucellus. The embryo continues to divide to form a **plumule** (embryonic shoot), **radicle** (embryonic root) and one or two **cotyledons**. Legume embryos develop two cotyledons that swell with starch, protein and fat. In cereals the single cotyledon is thin and is not a store; it absorbs nutrients from the endosperm.

The changes to the embryo occur inside the ovule as development is internal in the same way as it is in mammals. As the embryo gets larger, the nucellus is squashed and the integuments form the seed coat or **testa**. After fertilisation, the ovule becomes a seed and the ovary becomes a fruit. The fruit wall or pericarp enlarges to accommodate the growing seeds. The attachment to the ovule now reabsorbs water so that each seed dries out. The pericarp may retain water and become fleshy or it may also dry out. An example of a fleshy fruit is an orange; the fruits of *Capsella bursa-pastoris* are dry.

Seeds of *C. bursa-pastoris* are released when the fruit breaks open. When a seed breaks away from the inside of the ovule, it leaves a scar on the testa. Fruits are the remains of ovaries and have two scars – one where they were attached to the rest of the plant and one where the stigma was attached.

Genetic consequences of pollination

Sexual reproduction in flowering plants involves fusion of gametes that have been produced by meiosis. Self-pollination leads to self-fertilisation where the gametes come from the same plant. There will be some variation because meiosis 'shuffles' the genes and it is not the same meiosis that produces the pollen grain nuclei and the female gamete nucleus. Self-fertilisation is the most extreme form of **inbreeding**. If a plant has a potentially harmful recessive allele, there is a high probability that the offspring will be homozygous recessive for this allele.

If a flower has been self-pollinated then it is likely that all the seeds in the fruit will show very little variation as all the male gametes originated from the same parent. In cross-pollination the pollen grains that brought the male nuclei could have come from several or many plants and therefore within each fruit there could be a very wide range of genotypes. All the seeds have the same genes, but there could be a wide range of alleles for some or all of those genes. Remember that when we looked at the genetics of plants, we only considered one gene with two alleles when we crossed two varieties such as tall and dwarf pea plants. This meant that in this cross all the pollen grains carry a single allele. But if you consider many different genes the potential for variation within the seeds in one fruit is large. This is **outbreeding** which increases the amount of variation in the offspring generation.

Summary questions

1 Explain the difference between pollination and fertilisation.

2 Describe the events that occur in the ovule between pollination and seed dispersal.

3 Explain the significance of the double fertilisation that occurs in flowering plants.

4 State the functions of the following structures in the ovary between pollination and fertilisation: integuments, micropyle, pollen tube, tube nucleus, generative nucleus.

5 Make a table to compare asexual and sexual reproduction in flowering plants. Remember to include similarities as well as differences.

∞ Link

The summary section on pages 174 and 175 includes some exercises that will help you to compare internal fertilisation and internal development in flowering plants and mammals.

Figure 1.5.2 *Seeds of* C. bursa-pastoris *with fruits*

☑ Study focus

Look at tomatoes, oranges, limes and avocados to see that they each have two scars. They are fruits.

∞ Link

Plants have a gene that controls the production of chlorophyll. Answer Question 6 on page 156 to see the effect of inbreeding on the expression of this allele in the phenotype.

☑ Study focus

Before you start your table in answer to Question 5, make a list of the features to compare. Do not forget to include the features as the first column.

1.6 Plant reproduction summary

Learning outcomes

On completion of this section, you should be able to:

- sequence the events in sexual reproduction in flowering plants
- compare asexual reproduction with sexual reproduction
- use your knowledge and understanding of pollination to describe and explain the adaptations of local species.

 Study focus

Don't forget that you need to use mathematics in your biology course. This is an example of a topic where you need to apply your mathematical knowledge.

 Study focus

Before you start making your table for Question 6, make a list of the features of asexual and sexual reproduction. You might also make a spider diagram or other graphic organiser to help organise your knowledge of this topic.

The tasks and questions in this section will help you organise your knowledge and test your ability to compare sexual reproduction in flowering plants with sexual reproduction in humans.

1 Make a table to show the different groups of organisms described in this chapter and the ways in which they reproduce. You will need to do some research to find out how the species mentioned on pages 144–45 reproduce sexually.

Generation time (G) is defined as the length of time from one generation to the next. The mean generation time is calculated using the following formula:

$$G = \frac{t}{n} \qquad \text{where } t = \text{time and } n = \text{number of generations}$$

The table shows the rate of division of some species of bacteria.

Bacterium	Number of divisions in 24 hours	Generation time/min
Streptococcus lactis	55	
Escherichia coli	72	
Staphylococcus aureus	48	
Mycobacterium tuberculosis	18	

2 Copy and complete the table by calculating the generation time for each bacterium in minutes.

A typical medium for tissue culture contains the following components:

sucrose; amino acids; vitamins; nitrate ions, potassium ions, phosphate ions and trace quantities of iron ions and copper ions; an auxin and a cytokinin.

3 Find out why the components are necessary. Make a table to show these components and explain why they are required by the plant cells.

4 Find out how the following flowers are pollinated, list their main pollinator(s) and describe the adaptations that they have for attracting their pollinators:

Yucca, Crotalaria, Aristolochia, Cocos nucifera, Melicocca bijuga, Ruellia tuberosa, Bougainvillea, Euphorbia pulcherimma, Ipomoea

Present your findings as a poster or other form of presentation.

5 Find photomicrographs of pollen formation and embryo sac formation. Print the photographs and arrange them in the correct sequence. Annotate the photographs with information about meiosis and the development of pollen grains and embryo sacs.

You can use these photographs and notes to help with your revision by matching them together.

6 Make a table to compare asexual reproduction with sexual reproduction in plants, animals and microorganisms.

This is a list of events that occur during sexual reproduction in flowering plants (A to X)

7   Put these events in the correct sequence for a plant that shows protandry. The first one has been done for you – it is A.

Event during sexual reproduction	Letter		Event during sexual reproduction	Letter
flower bud opens	A		cell divides by mitosis three times	M
pollen mother cell divides by meiosis	B		ovum, synergids and antipodal cells form	N
insect pollinator visits flower	C		pollen tube reaches micropyle	O
pollen grains form	D		two nuclei move to centre of embryo sac	P
3 of 4 cells degenerate	E		male nucleus fuses with ovum	Q
pollen nucleus divides by mitosis	F		pollinator deposits pollen on stigma	R
generative and tube nuclei form	G		pollen grain germinates	S
pollen sacs ripens	H		pollen tube grows through style	T
dehiscence of anther	I		generative nucleus divides by mitosis	U
embryo sac mother cell divides by meiosis	J		tip of pollen tube breaks open	V
zygote forms	K		male nucleus fuses with polar nuclei	W
endosperm nucleus forms	L		pollen grain is compatible with stigma	X

To understand some of the genetics that was in Module 2, you needed to know something about reproduction in plants and animals. Now that you have covered Module 3 you can apply the knowledge from this module to genetics.

Self-incompatibility is determined by multiple alleles at the S locus. Pollen grains express one of the two alleles present in the pollen mother cell.

The table shows possible pairings between plants with five different genotypes. A minus sign indicates that no fertilisations are possible. In the others, either half ($\frac{1}{2}$) or all of the pollen is capable of germinating, growing pollen tubes and fertilising the ovules.

8   Copy and complete the table by writing '$\frac{1}{2}$' or 'all' where appropriate.

Genotype of male parent	Genotype of female parent				
	s1s2	s1s3	s2s3	s2s4	s3s4
s1s2	–				
s1s3		–			
s2s3			–		
s2s4				–	
s3s4					–

9   Explain why some pollen grains can germinate and fertilise ovules, but others cannot.

10 Use the results in your table to explain the advantage of self-incompatibility.

1 a With reference to named examples, describe TWO natural methods of asexual reproduction.
  b Explain why vegetative propagation is used in agriculture and horticulture.
  c Discuss the advantages and disadvantages of growing large areas with crops, such as ginger, Irish potatoes and sweet potatoes, that are produced by vegetative reproduction.

2 Some species, such as *Salvia roemeriana* from Mexico, have some flowers that open and are cross-pollinated by insect pollinators and others that never open.
  a Discuss the benefits of having two types of flower on the same plant.

  The periwinkle, *Catharanthus roseus*, originates from Madagascar but has been introduced throughout the tropics including the Caribbean. It has brightly coloured flowers.
  b Describe THREE ways apart from flower colour, in which plants, such as the periwinkle, attract insect pollinators.
  c The diploid number of *C. roseus* is 16. State the number of chromosomes in the following cells in an ovule after fertilisation: zygote, endosperm cell, antipodal cell, nucellar cell, integument cell.
  d Explain why fertilisation in flowering plants, such as *C. roseus*, is described as *internal*.
  e For each of the cells named in **c**, state the percentage contribution of the male parent and female parent to each nucleus following cross-fertilisation.

3 a Compare the structure of a pollen grain with the structure of a mature embryo sac.
  b Outline the ways in which cross-pollination may be achieved in flowering plants.
  c Explain why the seeds of a natural outbreeding plant, such as *Zea mays*, show more variation than the seeds of a natural inbreeder, such as the groundnut *Arachis hypogoea*.

4 a Make a labelled drawing of a cross section of an anther.
  b Describe how pollen mother cells give rise to pollen grains.
  c Describe the changes that occur in the ovule of a flowering plant between fertilisation and production of a mature seed.

5 a Explain the differences between the following pairs:
  i pollination and fertilisation;
  ii ovule and ovary;
  iii ovule and female gamete;
  iv seed and fruit.
  b Describe TWO methods to ensure that cross-pollination occurs.
  c Suggest how you would investigate the effects of inbreeding on a species of garden plant.

6 The pickerelweed, *Pontederia cordata*, is an aquatic plant that lives in shallow water on the edges of ponds and lakes. Albino seedlings of this species have been found. During an investigation plants were collected from the wild and cross-pollinated. This was continued for two generations. Among some of the crosses albino seedlings appeared as reported in the table. In each case the researchers deduced the genotype of the parental plants.

Cross	Genotype	Number of observed plants	
		green	albino
1	Aa × AA	149	0
2	Aa × Aa	70	22
3	Aa × Aa	65	28
4	AA × Aa	109	0

  a Explain how the researchers deduced the genotypes from the results of each cross. (You may use genetic diagrams to help your answers.)
  b Explain why it is not possible to carry out a test cross with these pickerelweed plants.
  c Discuss THREE ways in which cross-fertilisation between flowering plants is promoted.

7 Callus tissue is undifferentiated tissue which forms when explants of plant tissue are placed into a culture medium. An investigation was carried out into the effects of plant hormones on callus tissue. The callus tissue was prepared from leaves of tobacco, *Nicotiana tabacum*, and cultured on solid media containing different concentrations of two plant hormones, the auxin IAA and kinetin, a cytokinin. The results are shown below.

Treatment	Concentration of plant hormones/ mg dm⁻³		Effect of plant hormones on growth of callus tissue
	IAA	kinetin	
1	2.00	0.00	little or no growth
2	2.00	0.02	growth of roots
3	2.00	0.20	increased growth of callus with no differentiation
4	2.00	0.50	growth of shoots
5	0.00	0.20	little or no growth

a State THREE precautions that should be taken to ensure valid comparisons can be made between the treatments.

b Summarise the effects of the plant hormones on the growth of callus of *N. tabacum*.

c Explain how small quantities of callus tissue can be used to produce large numbers of plantlets of *N. tabacum*.

d Make THREE criticisms of the investigation as described above.

8 Tissue culture is a method of artificial propagation. One technique for carrying out tissue culture is to remove meristems from shoot tips and culture them.

a Describe the appearance of cells taken from a plant meristem.

b Suggest an advantage of using meristems in plant tissue culture.

c Outline TWO advantages and TWO disadvantages of using this technique to produce ornamental and crop plants.

d Suggest how raising plants in tissue culture makes it easy to genetically engineer plants.

☑ *Study focus*

Questions 6 and 8 ask about topics from other modules. This is unlikely to happen in the exam, but you should be aware that there are many connections between topics in Biology. Learning about flowering plant and human reproduction is useful background knowledge for Module 2. It is much easier to genetically modify cells in tissue culture than to modify whole plants, so this is an advantage of tissue culture which you can learn in the section on genetically modified organisms.

9 a Describe the changes that occur to the carpel of a flowering plant after fertilisation.

b Explain how both asexual reproduction and reproduction by seed can be successful strategies in different conditions in wild flowering plants.

c The world's climates may be changing. Outline the consequences of global climate change for plants that self-pollinate.

10 The diagram shows the growth of an embryo of the flowering plant Shepherd's purse, *Capsella bursa-pastoris*. The fruits of this species are non-fleshy and each contains several seeds.

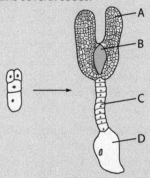

a Name the structures labelled A, B, C and D.

b Explain why the development of embryos in flowering plants, such as *C. bursa-pastoris*, is described as *internal*.

c Explain how the developing embryo receives the energy and nutrients it requires.

d Describe the changes that occur in the ovary wall as a fruit develops.

11 a Explain the differences between the following:
   i protandry and protogyny
   ii self-incompatibility and male sterility
   iii self-fertilisation and cross-fertilisation.

Double fertilisation occurs in the life cycle of flowering plants.

b Make a labelled diagram of a carpel at the time of fertilisation. Indicate on your diagram the structures that will fuse together at fertilisation.

c Explain the significance of double fertilisation.

3 Reproductive biology

2.1 Female reproductive system

Learning outcomes

On completion of this section, you should be able to:

- describe the structure of the human female reproductive system
- state the functions of the ovaries, oviducts, uterus, cervix and vagina.

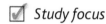 Study focus

There are several different oestrogen hormones. The most important is oestradiol.

In common with other mammals we have internal fertilisation and internal development. Unlike other mammals human females are fertile throughout the year. The cycle of changes that occurs in the reproductive organs of mammals is known as the oestrus cycle. In female humans this cycle occurs on average once a month and is known as the **menstrual cycle**. (The word *menstrual* is derived from a Greek word for moon and the Latin word for month.) The female system is coordinated so that the activities of **ovaries** and **uterus** are synchronised.

The functions of the female reproductive system are to:

- produce female gametes (secondary oocytes) in the ovaries
- transfer female gametes from the ovaries to the site of fertilisation in the oviduct
- provide a site for internal development of the embryo and then foetus for the nine months of the **gestation period** between fertilisation and birth
- secrete the female hormones, **oestrogens** and **progesterone**, which are steroids produced from cholesterol.

The table outlines the structure and functions of each organ labelled in Figure 2.1.1.

Organ	Structure	Functions
ovaries	pair of ovaries situated either side of uterus with outer germinal epithelium surrounding cortex and medulla	▪ produces the secondary oocytes ▪ secretes oestrogens, e.g. **oestradiol** ▪ secretes progesterone
oviducts (Fallopian tubes)	muscular tubes lined with ciliated epithelium – lead from fimbriae – feathery tissue surrounding ovary – to uterus	▪ transports secondary oocytes to uterus ▪ site of fertilisation ▪ transports embryo to uterus
uterus	inner lining is **endometrium** – rich in glands and blood vessels outer muscular layer is **myometrium**	▪ endometrium is site of development of embryo/foetus ▪ myometrium contracts during birth
cervix	neck of the uterus that separates uterus from vagina; consists of ring of muscle	▪ secretes mucus – consistency changes during each month ▪ cervix appears to be closed except around time of ovulation ▪ circular muscle at base of uterus retains contents during pregnancy ▪ plug of antibacterial mucus during pregnancy reduces infection ▪ muscle relaxes during birth so cervix dilates
vagina	muscular tube with folded inner lining	▪ epithelium secretes mucus ▪ bacteria produce lactic acid to provide acidic environment (pH 3.8 – 4.5) to prevent growth of other microorganisms ▪ site of deposition of semen during intercourse ▪ birth canal
vulva	external genitalia including clitoris and labia	▪ many sensory receptors for arousal during intercourse

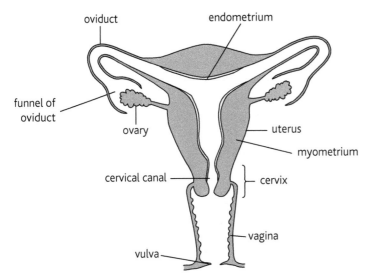

Figure 2.1.1 The human female reproductive system. The structures labelled are primary sexual characteristics and their functions are described in the table.

You may have to make a drawing from a prepared slide of a section through the ovary of a small mammal. Beware of comparing what you see in slides with diagrams as they often show all the stages of the ovarian cycle. You will not see these when you look at a slide.

After ovulation the cells of the Graafian follicle that remain in the ovary form the corpus luteum.

The table shows the functions of the regions of the ovary.

Region	Functions
germinal epithelium	▪ cells lining outside of ovary divide by mitosis (before birth) to form oogonia surrounded by follicle cells
cortex	▪ outer region contains primary follicles
Graafian follicle	▪ secondary oocyte develops within the Graafian follicle ▪ outer layer is the theca that secretes oestradiol ▪ fluid filled antrum creates pressure to burst follicle at ovulation
stroma	▪ forms theca of maturing follicles ▪ contains connective tissue and blood vessels blood vessels: ▪ supply nutrients to developing follicles and cholesterol for oestrogen and progesterone synthesis ▪ take away oestradiol from theca and oestradiol and progesterone from corpus luteum
corpus luteum	▪ secretes oestradiol and progesterone

The primary follicles develop before birth, so at birth there is a full complement of follicles that will last a woman the length of her reproductive life. If you answer Question 10 on page 177 you will see how many there are and what happens to the number over time.

Study focus

This photomicrograph shows the ovary at one stage of the menstrual cycle. Search for images that show other stages, or look for diagrams that show all the stages.

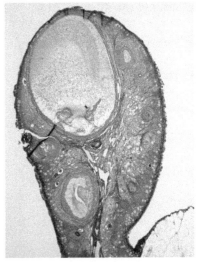

Figure 2.1.2 A photomicrograph of an ovary showing a Graafian follicle

Summary questions

1 Outline the roles of the following in the human reproductive system: ovary, oviduct, uterus, cervix and vagina.

2 Find a diagram or diagrams that show the changes that occur within the ovary during the ovarian cycle. Make a series of simple diagrams of the ovary to show the changes that occur during a cycle. Your diagrams should show the development of a follicle, ovulation and the development and degeneration of the corpus luteum.

3 Explain what happens in the development of the follicle.

4 Describe the structure of: **i** an ovary; **ii** the uterus. Outline the functions of the different tissues in each organ.

5 Explain why it is important to synchronise the changes in the ovary with changes in the uterus.

2.2 Male reproductive system

Learning outcomes

On completion of this section, you should be able to:

- describe the structure of the human male reproductive system
- state the functions of the testes, epididymis, vas deferens, prostate gland, seminal vesicles, penis, urethra.

The functions of the male reproductive system are to:

- produce very large numbers of male gametes which are known as spermatozoa (singular spermatozoon), but which we will call sperm or sperm cells for short
- produce a liquid medium, semen, for the delivery of sperm into the vagina
- secrete the male hormone, testosterone, which is a steroid produced from cholesterol.

The production of sperm cells does not involve the synchronisation of two organs as in the female. Instead there is the control of the production process in the testes so that millions upon millions of sperm are available throughout reproductive life.

The table outlines the structure and functions of the organs of the male reproductive system.

Organ	Functions
testis (plural testes)	• produces the male gametes inside **seminiferous tubules** • secretes testosterone
epididymis	• collects sperm from seminiferous tubules and reabsorbs fluid to concentrate them • stores sperm • site of maturation of sperm
scrotum	• sac containing the testes below the abdominal cavity where temperature is 2–3 °C cooler than body temperature
vas deferens	• muscular tube that moves sperm by **peristalsis** during intercourse
Cowper's gland	• secretes fluid that lubricates, cleans and neutralises urethra before and during ejaculation
prostate gland	• contributes to semen by secreting mucus and chemicals that activate sperm
seminal vesicle	• secretes seminal fluid as part of semen that contains glycoproteins that surround the surface of sperm • secretes fructose to provide energy for motility of sperm • seminal fluid is alkaline to neutralise contents of vagina to protect sperm
urethra	• contractile tube that moves semen through the penis
glans (penis)	• fills with blood to become erect • inserted into vagina during intercourse • sensory receptors at the tip for arousal during intercourse

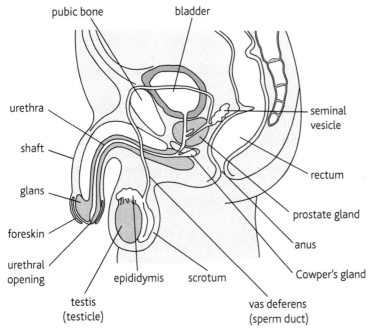

Figure 2.2.1 *The human male reproductive system. The functions of the primary sexual characteristics are described in the table.*

You may have to make a drawing from a prepared slide of a section through the testis of a small mammal.

The table shows the functions of the regions that are in the testis.

Region	Functions
tunica albuginea	■ outer layer of connective tissue
seminiferous tubules	■ site of sperm production ■ site of meiosis of diploid cells to form haploid cells ■ differentiation of haploid cells to form sperm ■ tubules connect together to form a duct into the epididymis
germinal epithelium (lining of seminiferous tubules)	■ cells divide by mitosis to form spermatogonia – from puberty throughout reproductive life
interstitial cells	■ secrete testosterone
blood vessels	■ provide cholesterol for testosterone synthesis and other nutrients for sperm production ■ remove testosterone to distribute around the body

Sperm cells do not develop properly at body temperature which is why the testes are held within the scrotum. Blood vessels flowing in and out of the scrotum flow close together so heat transfers from blood flowing into the scrotum to blood flowing back into the abdomen. This reduces the heat loss from the scrotum.

☑ *Study focus*

Search for photomicrographs, electron micrographs and drawings of seminiferous tubules to see the different stages of spermatogenesis and the Sertoli cells. These will help you with Question 2.

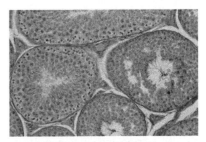

Figure 2.2.2 *A photomicrograph of a testis showing seminiferous tubules*

Summary questions

1 Outline the roles of the following in the human reproductive system: testis, epididymis, scrotum, vas deferens, seminal vesicle, prostate gland, urethra and penis.

2 Make a drawing of a sector of a seminiferous tubule showing the following: Sertoli cell, spermatogonia, primary spermatocytes, secondary spermatocytes, spermatids and spermatozoa.

3 Suggest the nutrients that are required for the production of sperm cells. Explain why you think each nutrient is required.

4 Explain why and how the temperature of the testes is kept several degrees below body temperature.

2.3   Gametogenesis

Learning outcomes

On completion of this section, you should be able to:

■ define the terms *gametogenesis*, *spermatogenesis* and *oogenesis*

■ describe the process of spermatogenesis to produce mature sperm

■ discuss the process of oogenesis to produce secondary oocytes

■ understand the difference between a secondary oocyte and an ovum

■ explain how the human gametes are adapted to their functions.

⮾ Link

When answering Question 3 make sure you include similarities in nuclear division and differences in cytoplasmic division.

Gametogenesis is the process by which gametes are produced. In humans, as in all vertebrates, this involves meiosis.

In the human life cycle gametes are produced by meiosis and are haploid. This is to ensure that each generation is diploid and the chromosome number does not increase when gametes fuse at fertilisation. On pages 82 and 84 we were concerned with the details of the *nuclear* division. When considering the production of human gametes we are concerned with the *cytoplasmic* division as well. **Spermatogenesis** is the production of sperm that occurs in the seminiferous tubules. Here the cytoplasm divides equally to give four haploid cells all of the same size. These cells then differentiate into mature sperm cells. **Oogenesis** is the production of an ovum. The nucleus divides in the same way as in spermatogenesis, but the cytoplasm divides unequally. This results in one large cell that may eventually become the ovum and three tiny cells that each contains a nucleus and very little cytoplasm.

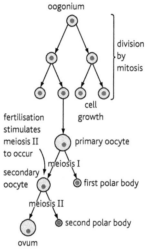

Figure 2.3.1 *Oogenesis. Note the unequal division of cytoplasm.*

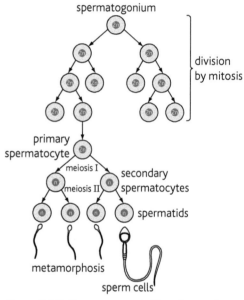

Figure 2.3.2 *Spermatogenesis. Note the equal division of the cytoplasm to give four haploid sperm cells. One sperm cell is enlarged to show its structure.*

Spermatogenesis begins at puberty. It is completed in the testis and takes about three months to produce mature sperm from a spermatogonium. About three million spermatogonia start the process each day. Oogenesis begins before the birth of a girl, but is not completed until half way through a menstrual cycle.

Both cells show the principle of structure related to function.

The functions of the sperm are to:

▪ deliver a haploid nucleus, with a set of paternal chromosomes, to the female gamete

▪ restore the diploid number

▪ increase genetic variation

▪ stimulate meiosis II in the secondary oocyte

▪ determine the gender of the next generation.

The functions of an oocyte are to:

▪ provide a haploid nucleus with a set of maternal chromosomes

▪ increase genetic variation

▪ provide energy supply for developing embryo.

Feature	Human sperm cell	Human secondary oocyte
linear dimension	length = 60 µm	width = 140 µm
overall structure	head piece (**acrosome** and nucleus), mid piece (axial filament and mitochondria), flagellum	spherical cell surrounded by zona pellucida and follicle cells
motility	swims using lashing movement of flagellum	some very limited motility
nucleus	haploid (23) – contains highly condensed DNA and histones to reduce the mass of the sperm.	haploid (23)
cell membrane	glycoproteins complementary to proteins on zona pellucida and oocyte membrane	microvilli to absorb nutrients;
cytoplasm	very little – to reduce size	contains organelles such as RER, SER, Golgi, mitochondria
energy store	very little – most energy provided by fructose from seminal vesicle	lipid droplets in cytoplasm
acrosome	present – contains enzymes, e.g. hyaluronidase and proteases for digesting pathway to oocyte	absent
mitochondria	arranged spirally around axial filament, to provide energy for swimming	many mitochondria to provide energy for development after fertilisation
centriole	one below nucleus – forms the microtubules of flagellum and forms centrioles of zygote	none – centrioles breakdown during oogenesis

Secondary oocyte

The oocyte is moved along the oviduct so does not have any structures involved with motility. After fertilisation all the energy for cell division comes from energy stores within the oocyte cytoplasm. The cell also needs supplies of lipids and proteins to make the new cell membranes for the many new cells that form. These sustain the embryo until it has implanted in the endometrium and absorbs nutrients from the mother's blood.

- Cell surface membrane has proteins that bind to proteins on sperm cells; also forms microvilli to absorb nutrients from the follicle cells.
- Cortical granules are lysosomes made by the Golgi body and contain substances released at fertilisation.
- Smooth endoplasmic reticulum to make lipids for new membrane formed after fertilisation.
- Rough endoplasmic reticulum to make proteins required immediately after fertilisation using mRNA produced by transcription before ovulation.
- Mitochondria provide energy for processes that occur after fertilisation.
- Lipid droplets are an energy reserve.
- First polar body is one of the haploid nuclei formed during meiosis I which contains a very small quantity of cytoplasm. It sometimes divides again by meiosis II.
- Zona pellucida is a region of glycoprotein secreted by follicle cells.
- Follicle cells form the corona radiata.

✓ Study focus

Secondary oocyte or ovum? The cell that is released from the ovary is a secondary oocyte as it has not completed meiosis. Fertilisation stimulates the completion of meiosis II followed by an unequal cytokinesis to form the tiny second polar body and the much larger ovum. The cell is called an ovum even though it has the sperm pronucleus inside it.

Summary question

1 Define the terms *gametogenesis*, *spermatogenesis* and *oogenesis*.

2 Explain the role of meiosis in gametogenesis in humans.

3 Make a table to compare oogenesis and spermatogenesis. Remember to include similarities as well as differences.

4 a Make large, labelled diagrams of a sperm cell and a secondary oocyte.
  b Explain how human sperm and secondary oocytes are adapted for their functions.

5 Explain how gametogenesis provides for genetic variation amongst the next generation.

Learning outcomes

On completion of this section, you should be able to:

- state where fertilisation occurs
- describe the process of fertilisation
- describe the early development of the embryo
- state where implantation occurs
- explain the process of implantation
- explain the biological principles of contraception.

Did you know?

Oocytes survive for about 12–24 hours after ovulation; sperm cells survive for up to 48 hours after ejaculation.

Did you know?

Men produce over 100 million sperm a day. One ejaculation contains 150 to 300 million sperm. All those sperm to fertilise one female gamete!

Did you know?

Sperm mitochondria are targeted by proteins in the cytoplasm of the embryo for destruction. As a consequence all mitochondria in the embryo are derived from those in the oocyte.

The Graafian follicle swells and protrudes from the side of the ovary. The pressure of the liquid inside increases and this causes the follicle to burst with the release of the oocyte and some surrounding follicle cells. This release of the oocyte and follicle cells is **ovulation**. If sexual intercourse occurs about the time of ovulation then there is a chance that fertilisation will occur. This period of time is the **fertile period**. Usually one ovary releases an oocyte during each menstrual cycle.

During intercourse the male inserts his penis into the woman's vagina. The stimulation of the penis results in contraction of the vas deferens from each testis so sperm are moved by peristalsis to the urethra. The seminal vesicle and the prostate gland release fluids that mix with the sperm to form **semen**. By further contractions the semen is moved along the urethra, through the penis and into the vagina.

The vagina is a hostile environment for sperm. The pH is low and this tends to kill sperm. The semen coagulates immediately after ejaculation and then liquefies 5 to 20 minutes later. They collect as a mass and move together through the cervix. Contractions of the cervix, uterus and oviduct help the sperm on their way towards the site of fertilisation at the ovarian end of an oviduct. During the fertile period the cervical mucus forms channels which groups of sperm swim through.

Sperm cells are not able to fertilise an oocyte until they have been through the process of **capacitation**. As they pass through the uterus and oviduct changes occur to the surface of the sperm. This takes about seven hours and involves the removal of the glycoproteins from the outer surface of the sperm. The cell membranes also become more permeable to calcium ions that increase their motility and help the release of enzymes from the acrosome.

Some sperm enter the oviducts and some will enter the oviduct with an oocyte. Sperm cells cluster around the oocyte.

Fertilisation

On contact with follicle cells, sperm cells are stimulated to release the enzymes in the acrosomes. This is known as the **acrosome reaction**. The acrosome swells and enzymes are released. These enzymes digest a pathway through the follicle cells to the cell surface membrane of the oocyte. Hyaluronidase digests hyaluronic acid between follicle cells and protease digests the proteins in the zona pellucida. The membranes of the two cells fuse and the whole sperm cell is taken into the oocyte cytoplasm. This immediately triggers changes in the oocyte with the cortical reaction. The contents of the cortical vesicles are released by exocytosis through the cell surface membrane to lift the zona pellucida and make it impenetrable to other sperm. This prevents **polyspermy** which is the entry of any more sperm.

Fertilisation triggers the final division of meiosis in the secondary oocyte. The oocyte nucleus divides and one of the new nuclei is extruded to form the second polar body (see page 162). The other nucleus is the **female pronucleus**. The nucleus from the sperm enlarges with the chromatin becoming less tightly coiled. This forms the **male pronucleus**. The nuclear membranes of the two pronuclei break down and the chromosomes assemble on the equatorial plate at metaphase of the first mitosis of the new diploid cell – the zygote.

The zygote divides into two cells to form a two-celled embryo. The two nuclei divide again by mitosis and the two cells divide by cytokinesis to form a four-celled embryo. This continues until the embryo becomes a hollow ball of cells, known as a **blastula**. Each cell is called a **blastomere**. Further growth occurs with the formation of:

- an inner cell mass that grows into the embryo and then into the foetus, and
- an outer layer of cells forming a **trophoblast** that later forms the placenta.

Six to nine days after ovulation the **blastocyst** embeds into the lining of the endometrium. The cells of the trophoblast form extensions from the surface of the blastocyst into the endometrium to increase the surface area for absorption of nutrients, water and oxygen. These are **trophoblastic villi** which protrude into small blood spaces or **lacunae** in the endometrium.

Contraception

Contraception literally means preventing conception. This involves preventing fertilisation of oocytes by sperm. The biological principles are summarised in the table.

Type of contraception	Biological principle	Method	Efficiency*
Sterilisation			
vasectomy	no sperm in semen	each vas deferens is cut and tied so sperm cannot pass from testis to urethra	0
tubal ligation	no oocyte in oviduct	oviducts are cut and tied so oocytes cannot pass along oviducts	0
Barrier			
condom (sheath)	sperm do not enter vagina during intercourse	condom is placed over penis before entry into vagina	3
femidom		femidom is placed into vagina before entry of the penis	3
diaphragm/cap	sperm do not enter cervix during intercourse	diaphragm is placed over entry to cervix	3–15
Contraceptive pills	no ovulation	pills contain one or both ovarian hormones to prevent ovulation	0–2

* Number of pregnancies per 100 women per year

Contraceptive pills are made from synthetic oestrogens and progestins to prevent ovulation. There are several types of contraceptive pill:

- the combined pill, which contains oestrogen and progestins; these act by negative feedback to inhibit secretion of GnRH, FSH and the mid cycle surge of LH
- the progestin-only pill (also known as the 'mini pill'); this may prevent ovulation but also causes a thickening of cervical mucus to reduce chances that sperm enter the uterus.

The pills are taken for 20 days and then for five days no hormones are taken to allow menstruation to occur.

The hormones are also delivered as a skin patch, an implant under the skin and may also be injected into muscle for slow release.

Did you know?

The trophoblast secretes hCG which is detected in pregnancy tests – even before the time when menstruation should occur.

☑ *Study focus*

The methods described here prevent fertilisation. There are other methods of contraception that prevent an embryo implanting in the uterus, such as the 'morning-after pill' and the coil.

Summary questions

1 Define the following terms: *fertilisation, implantation, contraception* and *sterilisation*.

2 Explain what happens to a sperm cell during capacitation.

3 Describe the events that occur at fertilisation.

The zygote divides many times to form a multicellular embryo as it moves along the oviduct into the uterus. This implants into the endometrium. The embryo continues to divide to produce new cells and some of these form tissues that will not be part of the baby when it is born. These extra-embryonic tissues form the trophoblast at the start of pregnancy and later form part of the placenta.

The extra-embryonic tissues develop into the **placenta** and the **amnion**. Sometimes they are called extra-embryonic membranes, but do not think they are cell membranes, they are layers of cells.

The placenta

The placenta is an organ formed from foetal tissue and from maternal tissue. The endometrium, blood in the sinuses and the cells lining these sinuses are maternal in origin, all the other tissues are foetal. The **chorion** is one of the extra-embryonic tissues and this grows to form the highly folded **chorionic villi** which increase the surface area for exchange of substances between maternal and foetal blood systems.

Maternal blood flows into the placenta from the **uterine artery**. This is oxygenated blood rich in nutrients. The blood flows into blood 'lakes' or sinuses which bathe the villi. Deoxygenated blood flows away from the placenta in the **uterine vein**.

Foetal blood flows from the heart along the aorta into arteries which carry blood to the legs. Here arteries branch to take this deoxygenated blood through the umbilical cord to the placenta. The blood flows through capillaries in the villi and returns as oxygenated blood in a vein that flows through the umbilical cord and then returns blood to the foetal heart. Only a small volume of blood flows through the foetal lungs as there is no point sending all the blood from the right ventricle as it will not be oxygenated.

The foetus has a different genotype from the mother. Without a barrier separating the two, the mother's immune system would reject the foetus. The placenta is a barrier to blood cells, many pathogens and some proteins.

Maternal blood in the uterine artery has a high blood pressure. If this entered the delicate foetal blood vessels they would burst. Pathogens, such as bacteria, some viruses and worms, are too large to cross the epithelium of the chorionic villi, but some viruses are able to cross. Many antibodies that circulate in the maternal blood are very large molecules that cannot cross. If they did cross some would be likely to do serious damage to foetal red blood cells. There are some smaller antibodies that cross the placenta to give the foetus immunity to diseases that the mother has had.

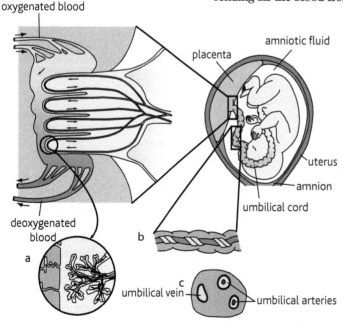

Figure 2.5.1 *The structure of the placenta showing: **a** detail of the placental villi; **b** the umbilical cord with its three blood vessels; **c** a cross section of the umbilical cord*

Exchange of gases

Foetal red blood cells contain a form of haemoglobin with two α-globins and two γ-globins. This binds oxygen more readily than the haemoglobin in the maternal red blood cells, so encouraging the diffusion of oxygen across the placenta, down the concentration gradient into the foetal blood. Carbon dioxide is in higher concentration in the foetal blood than in the maternal blood so it diffuses in the opposite direction.

Absorption of nutrients

Glucose is absorbed by facilitated diffusion. Some glucose is stored in the placenta as glycogen and much is used in respiration by the placenta and foetus. Amino acids are in a higher concentration in foetal blood than in maternal blood so they are absorbed by active transport. Ions and water soluble vitamins are also absorbed. Fats and fatty acids and fat soluble vitamins are probably absorbed by pinocytosis (see page 39). Antibodies are absorbed into vesicles which transport them across the epithelial cells to transfer them into foetal blood.

Water moves across the epithelial membrane by osmosis. If the foetal blood has a lower water potential than maternal blood then water will be absorbed. Water will move from foetal blood to maternal blood if the foetal blood has the higher water potential. After about 11 weeks the foetal kidneys start to function and urine is produced that is released into the amniotic cavity.

The foetal metabolism produces carbon dioxide which diffuses across the placenta; urea and bile pigments also diffuse through the phospholipid bilayers of the capillary and epithelial cells into the maternal blood.

The amnion

The amnion also forms from extra-embryonic tissues to surround the foetus. It is a thin, but tough sheet of tissue that is transparent; it secretes a fluid that has much the same composition as tissue fluid. This **amniotic fluid** fills the **amniotic cavity** that surrounds the foetus providing a sterile environment and a cushion for the foetus so it is not affected by mechanical damage. The fluid is derived from maternal blood plasma but with lower protein content. The fluid provides a constant temperature for the foetus, which is especially important if the mother is exposed to extremes of temperature.

The foetus is able to move about within the amniotic fluid which helps the development of the skeletal and muscular systems. The foetus absorbs some of the liquid through the skin which does not become tough and impermeable until about 20 weeks. The foetus drinks the amniotic fluid which helps development of the swallowing reflex. The liquid passes through the foetal gut and is absorbed; some of it is then urinated into the amniotic fluid. The foetus also starts breathing movements. The amniotic cavity and its fluid help to prepare the foetus for its independent life as a baby after birth.

The amnion breaks during the early stage of birth.

 Study focus

The placenta has four roles:

- separation of maternal and foetal circulatory systems
- exchange of substances between maternal blood and foetal blood
- secretion of hormones
- metabolism and energy storage.

 Study focus

Lennart Nilsson is a Swedish photographer who has specialised in human development. Search for his photos online.

⊖⊂ Link

Read pages 38 to 39 to remind yourself about exchanges across membranes. Cell membranes in the epithelium of the placental villi have carrier proteins for facilitated diffusion and active transport. The cells contain many mitochondria to provide energy for active transport.

Summary questions

1 Explain: **i** why the placenta is an organ, and; **ii** how the placenta acts as a gas exchange surface and as an excretory organ.

2 Explain how the placenta is adapted for exchange of substances.

3 Describe the pathway taken by blood in the foetus from the heart to the placenta and back.

4 Describe the roles of the amnion.

5 Explain two ways in which the foetus may change the composition of the amniotic fluid.

2.6 Hormonal control of reproduction

Learning outcomes

On completion of this section, you should be able to:

- define the term *hormone*
- name the hormones that regulate reproduction in humans
- explain how hormones regulate gametogenesis
- describe the role of hormones in the control of the menstrual cycle
- state the role of negative feedback in hormonal control of reproduction
- describe the role of the placenta in secreting hormones.

☑ Study focus

The term *gonadotrophic* means stimulating the growth of the gonads – the ovaries and testes.

∞ Link

Cells in target tissues and organs have receptors to detect hormones. See page 73 for information on the testosterone receptor.

☑ Study focus

The changes in the ovary and the uterus must be synchronised so that the uterus is ready to receive the embryo if fertilisation occurs.

☑ Study focus

Follow the changes in concentrations of the four hormones on these graphs.

The production of sperm is a continuous process so the secretion of hormones to control spermatogenesis is also continuous. Oocytes are produced at intervals of about a month in women who are not pregnant or breast feeding and not on the 'pill'. The changes that occur in the female reproductive systems are cyclic.

A **hormone** is a signalling molecule secreted by an endocrine organ. It is transported in the blood and influences the activity of target tissues and/or organs. The hormones that control the activity of the reproductive organs are:

- **gonadotrophin releasing hormone** (GnRH) released by the hypothalamus with the anterior pituitary gland as the target organ; it is a decapeptide with 10 amino acids
- **gonadotrophic hormones** released by the anterior pituitary gland influence the activity of the gonads – the testes and ovaries; **follicle stimulating hormone** (FSH) and **luteinising hormone** (LH) are glycoproteins.

In females the interaction between the hormones is complex as you can see in Figure 2.6.1.

There are receptors for GnRH, LH and FSH on the cell membranes of their target cells.

- **gonadal hormones** are steroids released by cells in the gonads: the ovaries secrete oestrogens and progesterone; the testes secrete testosterone. They stimulate gametogenesis, the development of primary and secondary sexual characteristics, and in females coordinate the activities of ovary and uterus. They function by interacting with receptors inside cells and 'switching on' the transcription of genes.

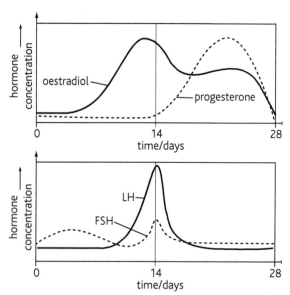

Figure 2.6.1 *This graph shows the changes in the concentration of FSH, LH, oestradiol and progesterone in the blood of a woman during a menstrual cycle*

Before puberty these hormones are secreted but at very low concentrations. The cells in the hypothalamus detect the hormones secreted by the gonads and maintain a low concentration by releasing low concentrations of GnRH. At puberty the secretion of GnRH increases so the concentration in the blood increases. This stimulates an increase in secretion of gonadotrophins by the anterior pituitary.

The table shows the details of hormones that you need to know.

Hormone	Site of production	Target organ/ tissue	Action
gonadotrophin releasing hormone (GnRH)	hypothalamus	pituitary gland	stimulates anterior pituitary to release FSH and LH
follicle stimulating hormone (FSH)	anterior pituitary gland	ovary	stimulates oogenesis
		testes	stimulates Sertoli cells to develop sperm cells
luteinising hormone (LH)	anterior pituitary gland	ovary follicle	stimulates release of secondary oocyte at ovulation
		testes interstitial cells	stimulates interstitial cells of testes to secrete testosterone
oestrogen	follicle in ovary	uterus	stimulates repair and growth of endometrium stimulates growth and development of primary and secondary sexual characteristics in females
	interstitial cells in testes	testes and epididymis	regulates spermatogenesis and sperm maturation in epididymis
progesterone	corpus luteum in ovary	uterus	maintains endometrium
	testes	seminiferous tubules	helps regulate sperm production increases sperm motility
testosterone	ovaries	brain	one of many factors in determining sex drive
	interstitial cells of testes	seminiferous tubules	stimulates spermatogenesis stimulates growth and development of primary and secondary sexual characteristics in males
inhibin	Sertoli cells in testis	hypothalamus and anterior pituitary	inhibits secretion of FSH from anterior pituitary
	follicle in ovary	anterior pituitary	inhibits secretion of FSH

Menstrual cycle

The changes that occur within the ovary and the uterus are collectively known as the menstrual cycle. The changes that occur in the ovary are known as the **ovarian cycle**; the changes in the uterus are known as the **uterine cycle**. These changes are summarised in the table.

Days	Hypothalamus	Anterior pituitary gland	Ovarian cycle	Uterine cycle
1–5	secretion of GnRH starts	secretion of FSH and LH starts	*follicular phase:* development of primary follicle starts secretion of oestradiol; one dominant follicle develops into Graafian follicle	*menstrual phase:* menstruation occurs
5–13	oestradiol inhibits secretion of GnRH	FSH secreted, but gradually inhibited by low concentrations of oestradiol surge of LH secretion		*proliferative phase:* repair and growth of endometrium
14	secretion of GnRH	FSH and LH secreted	*ovulation* – release of secondary oocyte	
15–18	progesterone inhibits GnRH secretion decrease in GnRH secretion	concentration of LH gradually decreases decreasing concentrations of oestradiol progesterone inhibits FSH and LH secretion	*luteal phase:* corpus luteum secretes progesterone, reduction in oestrogen secretion, corpus luteum degenerates, less progesterone secreted	*secretory phase:* glandular activity in endometrium maintains its thickness; decrease in progesterone concentration leads to breakdown of endometrium
18–24		secretion of FSH and LH not inhibited		
24–28				

Negative feedback

The secretion of hormones is carefully controlled so that the concentration in the blood stays within certain limits and does not keep increasing. This is achieved by the principle of **negative feedback**.

In males after puberty, the hypothalamus maintains the concentration of testosterone between 300 and $1000\,\mu g\,100\,cm^{-3}$ which is the set point. If the concentration rises:

- the hypothalamus reduces the secretion of GnRH
- the anterior pituitary reduces the secretion of LH.

In addition, the hormone inhibin secreted by Sertoli cells inhibits the secretion of FSH from the anterior pituitary.

If the concentration falls below the set point, then the hypothalamus increases the secretion of GnRH.

In females the interaction between the hormones is more complex. The changes during the menstrual cycle are as follows:

Days 1–5: low concentrations of progesterone and oestradiol; GnRH secreted by hypothalamus; FSH and LH secreted by anterior pituitary.

Days 6–13: FSH stimulates theca cells to secrete oestrogen; as concentration of oestrogen increases there is a negative feedback on hypothalamus and anterior pituitary so concentrations of GnRH and FSH decrease. But at the same time there is a positive feedback that causes the concentration of LH to increase.

Day 11: there is a surge in LH and FSH concentrations

Day 14: concentrations of FSH and LH peak; ovulation occurs and follicle cells form corpus luteum.

Days 15–28: LH stimulates corpus luteum to secrete oestradiol and progesterone; if fertilisation does not occur, progesterone exerts a negative feedback on the hypothalamus and anterior pituitary so that concentrations of FSH and LH decrease. As a result the concentration of progesterone decreases leading to menstruation. With very low concentrations of progesterone there is no inhibition of GnRH secretion and another cycle starts.

Secretion of hormones during pregnancy

Once the embryo has implanted in the endometrium, it must announce its arrival. The cells of the trophoblast secrete the hormone chorionic gonadotrophin which ensures the survival of the embryo by stimulating the corpus luteum in the ovary to continue to secrete progesterone. This stimulation continues beyond the end of the fourth week of the menstrual cycle so ensuring that the lining of the endometrium does not break down carrying away the implanted embryo. The secretion of chorionic gonadotrophin is taken over by the placenta as soon as it develops.

The placenta is an endocrine organ secreting a number of hormones as summarised in the table.

✓ *Study focus*

You should use the information in this section to draw feedback loops.

✓ *Study focus*

In this example, the set point is the desired concentration. You choose the set point when you adjust the temperature on the air conditioning or set the temperature for an oven.

Hormone	Target organ(s)	Function
chorionic gonadotrophin (often known as hCG with h standing for human)	corpus luteum in ovary	▪ stimulates continued secretion of progesterone to maintain pregnancy for first three months
progesterone	uterus	▪ maintains endometrium ▪ inhibits contraction of myometrium ▪ prevents menstruation
	hypothalamus and pituitary	▪ inhibits secretion of GnRH and FSH
	breasts	▪ development of milk glands
oestrogen	uterus	▪ stimulates growth of uterine wall ▪ sensitises myometrium to oxytocin that is secreted at birth to stimulate muscle contraction
	anterior pituitary and hypothalamus	▪ inhibits secretion of GnRH and FSH
	breasts	▪ stimulates growth of ducts

Summary questions

1 Define the following terms: *hormone*, *endocrine organ*, *target organ* and *negative feedback*.

2 Name the hormones that control reproduction in humans.

3 Outline the changes that occur in the ovary and uterus during one menstrual cycle.

4 Explain how hormones control gametogenesis in males and females.

5 Outline the roles of hormones in synchronising the activities of the ovary and uterus.

6 Explain how negative feedback is involved in controlling: **i** testosterone secretion; **ii** the menstrual cycle.

7 Describe the role of the placenta in secreting hormones.

8 Make a large copy of the graphs on page 168 and annotate them with information about where the hormones are secreted, where they act and what they stimulate. Draw a diagram with the same timeline on the *x*-axis to show changes in the uterus and the ovary.

Learning outcomes

On completion of this section, you should be able to:

- explain the importance of maternal nutrition during pregnancy

- outline the effects of drug abuse on foetal development including alcohol, nicotine, other legal and illegal drugs

- describe the effects of cigarette smoking on foetal development

- explain why some diseases pose health risks to unborn children.

✓ Study focus

You should read some advice given to pregnant women at antenatal clinics or talk to the staff at such a clinic. This will help you to understand the topics in this section.

A foetus interacts with the surrounding world entirely through the placenta. This organ acts as its lungs, gut and excretory system. Mothers should maintain a good diet, stay healthy and not take part in any risky behaviour, such as overusing legal drugs and abusing illegal drugs. Effects of maternal behaviour may affect the infant (up to one year after birth) or may be long-lasting and affect a person throughout their life.

Maternal nutrition

A woman's metabolism goes through many changes when pregnant. There is an increased demand for energy and nutrients to support the growth and metabolism of the uterus, placenta, breasts and foetus. During the first six months of pregnancy this is provided by the mother without any additional requirements from the diet; she does not need to 'eat for two'. She is able to make use of stored fat and stores of micronutrients, such as calcium and iron. She also becomes more efficient at absorbing nutrients from her diet. Towards the end of pregnancy it is necessary to increase the intake of energy and certain nutrients as shown in the table.

Nutrient	Function	Comments on intake during pregnancy
energy (provided by carbohydrates, fats and proteins)	required for growth and metabolism of uterus, placenta, foetus and breasts	energy intake should *not* increase until the last three months of pregnancy and then by about 800 kJ a day
protein	growth of tissues, such as blood in both mother and foetus and muscle in uterus and foetus	additional 6 g a day throughout pregnancy
calcium	growth of bones of foetus	no extra calcium needed as absorption from gut is more efficient
iron	haemoglobin in both mother and foetus	women at risk of iron-deficiency anaemia should take a supplement of iron with vitamin C

Women are advised to take a folic acid supplement to ensure that they have 400 µg per day for one month *before* becoming pregnant and for the first 12 weeks of pregnancy to reduce the risks of neural tube defects, such as spina bifida. In this condition the neural tube that goes on to develop into the brain and spinal cord does not close properly and is exposed at the surface of the body at birth.

∞ Link

Read about body mass index (BMI) on page 11.

It is not necessary to increase the food intake by very much during the last three months in order to gain sufficient energy. In fact, women who are overweight or obese should probably not increase at all and should consider losing weight during pregnancy. A woman of normal weight range (BMI of 18.5–24.9) should increase by 11–15 kg in body mass.

Pregnant women should take in sufficient water so the mother is not dehydrated and there is sufficient water for the amniotic fluid.

Drugs

A drug is any substance taken into the body that modifies or affects chemical reactions in the body. This includes medicines, legal drugs such as alcohol and nicotine, and illicit drugs, such as heroin and cocaine.

Alcohol crosses the placenta and enters the foetal blood circulation and is distributed throughout the foetus. This is at a time before the foetus is able to metabolise alcohol and so the effects are worse on the foetus than on the adult. Mothers who drink alcohol are increasing the risk of premature delivery. A very rare condition caused by heavy intake of alcohol during pregnancy (especially the first eight weeks) is **foetal alcohol syndrome** (FAS).

Nicotine is the drug in tobacco products, such as cigarettes. Like alcohol it crosses the placenta and enters the foetal circulation having the same effects as on adults:

- increases heart rate
- restricts blood flow to the extremities
- increases blood pressure
- increases stickiness of platelets so increasing risk of blood clots.

Nicotine reduces the flow of blood through the umbilical arteries and through the placenta.

Heroin and **cocaine** pose very serious health risks to unborn children. The drugs cross the placenta and the foetus becomes dependent on the drug in the same way as the mother. This means that after birth, the child shows withdrawal symptoms, such as severe shaking; cot deaths are more common amongst babies of women dependent on drugs, such as heroin and cocaine.

Cigarette smoking

Carbon monoxide in cigarette smoke combines irreversibly with haemoglobin to form **carboxyhaemoglobin**. This reduces the ability of haemoglobin to transport oxygen by up to 10 per cent and affects both the maternal blood and foetal blood. The combined effects of nicotine and carbon monoxide lead to the following:

- umbilical blood vessels are narrower
- placenta is smaller than normal
- restriction of blood through placenta may stimulate foetal heart to beat faster
- increased risks of miscarriage and premature birth
- increased risk of death of foetus or infant just before or after birth
- decreased mass at birth as a result of intra-uterine growth retardation (IUGR)
- decreased immunity to infection.

✓ *Study focus*

Pregnant women who have diets deficient in important nutrients, such as calcium, iron and folic acid, put the health of their children at risk. Try answering Question 6 on page 176 to work out the effect of such deficiencies on the developing foetus.

Summary questions

1 Explain why the following advice is given to women who are planning to become pregnant: eat a balanced diet; take folic acid tablets; stop smoking; stop drinking alcohol; do not increase energy intake until the last three months of pregnancy.

2 Find out about the effects of the drug thalidomide on the children of mothers who had taken the drug during pregnancy. Summarise your findings as a series of bullet points.

3 Find out the effects of foetal alcohol syndrome and make a list of them.

4 Summarise the effects of the following on foetal development: reduced energy intake, nutrient deficiencies, alcohol, cigarette smoking and illicit drug taking.

5 Pathogens, such as bacteria and viruses, cause disease. Find out which pathogens can cross the placenta and infect the foetus. What precautions can women take to prevent this?

2.8   Human reproduction summary

Link

Can you name any flowering plants and mammals that are fully aquatic? Look at page 150 for an example of a flowering plant.

The tasks and questions below will help you organise your knowledge of this subject area. Most flowering plants and mammals are terrestrial, they live on land. They have very different adaptations to reproduction and dispersal on land, but there are some aspects of their reproduction that are remarkably similar. There are some questions here to help you to make comparisons between sexual reproduction in flowering plants and sexual reproduction in humans and to understand the underlying principles that apply to both groups.

This is a list of events that occur during sexual reproduction in humans (A to X).

1 Put these events in the correct sequence. The first one has been done for you – it is A.

✅ *Study focus*

To help you follow the changes that occur during reproduction, draw a time line for a male and a female from the time of their conception to the time when they conceive a child.

Event during sexual reproduction	Letter	Event during sexual reproduction	Letter
germinal epithelium in testes produces spermatogonia by mitosis	A	germinal epithelium in ovaries produces oogonia	M
implantation	B	zygote formed	N
sperm stored in epididymis	C	2° spermatocytes divide to form spermatids	O
fertilisation	D	sperm pass through cervix and uterus	P
1° oocytes divide to form 2° oocytes	E	oogonia grow into 1° oocytes	Q
sexual intercourse – sperm ejaculated in semen	F	sperm enter oviduct	R
spermatids differentiate into spermatozoa	G	ovulation	S
spermatogonia grow into 1° spermatocytes	H	sexual intercourse – sperm deposited in vagina	T
zygote divides to form 2-celled embryo	I	1° spermatocytes divide to form 2° spermatocytes	U
internal development in the uterus	J	blastocyst forms	V
seminal vesicles, Cowper's gland and prostate gland release secretions	K	birth	W
embryo develops into a foetus	L		

✅ *Study focus*

You need to organise your sequence in two parts. The first part will show gametogenesis in males and females; the second part will show the changes that occur after fertilisation.

✅ *Study focus*

This sequence tells a story. There are many examples of processes that you can learn as if they are stories with a beginning, middle and end. Think of replication, transcription, translation, mitosis, meiosis. If you cannot tell the story, you will not be able to write about it in the exam.

2 Make a table to compare oogenesis and spermatogenesis.

Include the following headings for your rows: time of the life cycle when process starts; time of the life cycle when the process stops; number of gametes produced from each cell from each diploid cell; precise site of production within gonad; names of other cells involved in production of gametes; equal or unequal division of cytoplasm; length of time.

3 Identify each of the following hormones from these statements:

A Secreted by interstitial cells.

B Stimulates secretion of FSH.

C Made from cholesterol in the corpus luteum to maintain endometrium.

D Stimulates repair and growth of the endometrium after menstruation.

E Acts by negative feedback to inhibit secretion of LH from anterior pituitary after ovulation.

F Secreted by Sertoli cells to inhibit secretion of FSH.

4 The concentration of GnRH is kept within narrow limits in males. Draw a feedback loop to show how the concentration of GnRH is controlled by negative feedback.

5 a Explain how polyspermy is prevented.

b Explain why it is important to prevent polyspermy.

6 Make a flow chart to show the changes that occur after a sperm cell reaches the follicle cells surrounding an oocyte until the zygote divides to give a two-celled embryo.

7 Explain the biological principles involved with the following methods of contraception: vasectomy, tubal ligation, barrier methods and contraceptive pills.

8 Make a table to compare sexual reproduction in flowering plants and mammals. Include the following row headings for your table: site of meiosis; method of delivery of male gamete; type of fertilisation (single or double); names of structures that fuse together; site of fertilisation; product(s) of fertilisation; number of chromosome sets in products of fertilisation; site of internal development; method of providing nutrients to embryo.

9 Make diagrams to compare the role of meiosis in the production of a mature embryo sac from a megaspore mother cell in a flowering plant and meiosis from a primary oocyte to produce a mature human ovum.

10 Make a table to show the precise location of each of the following and its ploidy number in sweet potato $(2n = 90)$ and a human.

Sweet potato: megaspore mother cell, pollen mother cell, male gamete, antipodal cell, zygote, endosperm cell, meristematic cell.

Human: oogonium, primary oocyte, secondary oocyte (two locations), zygote, 8-celled embryo, trophoblast cell, spermatogonium, primary spermatocyte, secondary spermatocyte, spermatozoon (numerous sites).

11 Describe how the placenta acts as a life support system for the foetus. Compare the functions of the placenta with the ways in which a plant embryo is supported.

 Link

Read page 80 to remind yourself about the human life cycle before making your table.

1 a Compare the functions of the testes and ovaries in the male and female reproductive systems.

b Sperm cells go through the process of capacitation after ejaculation into the female reproductive tract. Describe what happens during capacitation.

c Contraceptives work in a variety of different ways. State **i** TWO methods that prevent sperm reaching an oocyte; **ii** TWO methods that prevent an oocyte reaching the site of fertilisation in the oviduct.

2 a Name the cells formed in oogenesis and spermatogenesis immediately after each of these processes:

mitosis, growth phase, meiosis I, meiosis II.

b Explain the significance of mitosis and meiosis in spermatogenesis.

c Describe the role of Sertoli cells in seminiferous tubules.

3 a Explain how negative feedback is involved in the control of the menstrual cycle.

b Deficiencies in the diet may adversely affect a woman's menstrual cycle. Explain how deficiencies in two named nutrients may cause the menstrual cycle to be interrupted.

c A secondary oocyte is released from an ovary at ovulation surrounded by a zona pellucida and follicle cells.

i Describe how a sperm cell penetrates these surrounding layers and enters the cytoplasm of an oocyte.

ii Explain how polyspermy is prevented.

4 a Describe the process of implantation.

b i State a site of secretion of the hormone human chorionic gonadotrophin (hCG).

ii Explain how the secretion of hCG causes menstrual cycles to stop.

c During pregnancy approximately 300 mg of iron are transported from the mother to the foetus. Under normal circumstances, pregnant women do not need to take iron supplements unless they become anaemic. Explain why

i iron is necessary for the foetus;

ii many pregnant women do not need to take iron supplements;

iii a foetus may suffer if the mother becomes anaemic.

5 a Describe the roles of the following during pregnancy: human chorionic gonadotrophin, corpus luteum, oestrogen and progesterone.

b Describe FOUR different ways in which substances cross the epithelial membranes of the chorionic villi in the placenta. Give an example of a substance that crosses in each way you describe.

c Explain why women are advised not to smoke and not to drink alcohol during pregnancy.

6 a State THREE functions of the amnion and amniotic fluid.

b State ONE similarity and ONE difference between the ways in which a human foetus and a plant embryo receive nutrients during their development.

c Explain how pregnant women should take care of their diet during pregnancy.

∞ Link

You can expect questions that ask you to compare the reproduction of flowering plants and humans. Make sure you prepare yourself for this type of question by doing Questions 8–11 in the Summary section on page 175.

7 a Describe and explain TWO ways in which the composition of the amniotic fluid changes during pregnancy.

b Describe the consequences for the development of a foetus if the mother consumes: **i** excessive quantities of alcohol; **ii** a diet deficient in vitamins during pregnancy.

c Mitochondrial DNA (mtDNA) is the loop of DNA found in each mitochondrion. Both sperm and oocytes have mitochondria but mitochondrial genes are only inherited from the oocyte. Explain how it is that mtDNA is inherited from the female parent only.

8 a Outline the roles of the following hormones in spermatogenesis: GnRH, LH and FSH.

b Explain how the concentrations of these hormones are controlled within narrow limits.

c A type of female contraceptive is Depo-Provera which is injected into muscle tissue every 12 weeks. Suggest:

i how this contraceptive prevents pregnancy

ii why one injection of this contraceptive can last for as long as 12 weeks

iii how the effectiveness of this contraceptive may be determined.

9 The graph shows changes in the concentrations of four hormones during a woman's menstrual cycle.

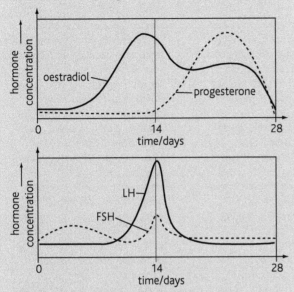

Figure 2.9.1 This graph shows the changes in the concentration of FSH, LH, oestradiol and progesterone in the blood of a woman during a menstrual cycle

a Describe the changes in concentrations of FSH and LH during the menstrual cycle.

b Explain the roles of FSH and LH during the first half of the menstrual cycle.

c Describe how the concentrations of the hormones change if the woman becomes pregnant. You may draw a sketch graph to help your answer.

10 The table shows the results of a study into the changes in numbers of oocytes in the ovaries during reproductive life.

Age/years	Mean number of oocytes per female
birth	712 000
7	468 635
14	402 067
21	175 700
28	166 231
35	89 145
42	39 874
49	9956
56	3450

a i Plot a graph to show the changes in the mean number of oocytes.

ii Calculate the percentage decrease in the mean number of oocytes between birth and age 42 years.

iii Suggest why there was a decrease in the numbers of oocytes.

b State the functions of the following structures in the female reproductive system:

i oviduct

ii uterus

iii cervix.

c A woman is advised to stop smoking cigarettes while she is pregnant. Outline the reasons for this advice.

11 a i Draw a labelled diagram to show the structure of a human sperm cell.

ii Explain how the sperm is adapted to its function.

b Describe how the structure of a human secondary oocyte differs from that of a sperm cell.

It is estimated that there are more than 20 million sperm in each 1 cm³ of semen. In women who are not pregnant or not taking the 'pill', one or two secondary oocytes are released each month.

c Explain why many more sperm are produced than oocytes.

12 a Describe how male gametes are produced in mammals (details of the stages of meiosis are not required).

b Explain how the production of male gametes in a flowering plant differs from the production of male gametes in mammals.

c Describe how male gametes reach the site of fertilisation in mammals and flowering plants.

Glossary

A

acrosome sac of enzymes on head of a sperm cell, derived from a lysosome.

activation energy energy that must be overcome before a reaction can proceed.

active site part of an enzyme molecule where the reaction takes place; has a complementary shape to the substrate.

active transport movement of molecules against a concentration gradient across a cell membrane using energy from respiration.

adenine a purine base, a component of DNA, RNA and ATP (adenosine triphosphate).

adventitious roots roots that grow from any structure other than the main (tap) root; they grow from leaves and stems.

allele an alternative form of a gene. There may be two alleles of a gene (e.g. A and a). Some genes have multiple alleles, e.g. the blood group gene which has three (I^A, I^B and I^O). The alleles of a gene have different DNA nucleotide sequences, but occupy the same locus on homologous chromosomes.

allopatric speciation formation of a new species from populations that are isolated from one another by geography; see sympatric speciation.

amino acid activation attachment of amino acids to specific tRNA molecules.

amnion extra-embryonic tissue that surrounds the foetus enclosing amniotic cavity filled with amniotic fluid.

amylopectin a polysaccharide that with amylose forms starch; composed of branched chain of α-glucose.

amylose a polysaccharide that with amylopectin forms starch; composed of unbranched chain of α-glucose in a helix.

anabolism reactions that build up larger biological molecules from smaller ones.

anaphase stage of nuclear division in which sister chromatids move apart (mitosis and meiosis II) and chromosomes move apart (meiosis I).

aneuploidy not all chromosomes are present in equal numbers a result of chromosome mutation involving loss or gain of individual chromosomes.

anther part of a stamen that has four pollen sacs which produce pollen grains.

antibiotic a compound produced by microorganism (or produced synthetically) to kill or inhibit the growth of bacteria.

antibiotic resistance ability of bacteria to survive use of an antibiotic; resistant bacteria have a mutant gene that gives ability to withstand or breakdown that antibiotic.

anticodon group of three bases on tRNA that pairs with codon on messenger RNA during translation.

antiparallel term used to describe the orientation of the two polynucleotides in DNA; given as 5'–3' and 3'–5', the numbers referring to the carbon atoms in deoxyribose.

aseptic technique term used to describe techniques used to ensure that microorganisms and cells in tissue culture are cultured in sterile conditions and not contaminated by unwanted viruses, bacteria or fungi.

asexual reproduction reproduction in which new individuals are formed from one parent without fusion of gametes, e.g. binary fission, budding, vegetative reproduction.

autosome any chromosome other than a sex chromosome; autosomes exist in homologous pairs that match exactly in appearance although they may have different alleles of the genes they carry.

B

bell-shaped curve term used to describe a normal distribution in a feature showing continuous variation in which the curve drawn through the blocks of a histogram is: symmetrical and the mean, mode and median all coincide.

binary fission form of asexual reproduction when a cell divides into two.

biological species population of interbreeding organisms that share features in common and reproduce sexually to produce fertile offspring; see morphological species.

bivalent a pair of homologous chromosomes as they associate together during prophase and metaphase of meiosis I.

blastomere one of the cells of a blastula.

blastula animal embryo that is a hollow ball of cells formed by repeated divisions of the zygote.

budding form of asexual reproduction when a small outgrowth enlarges and breaks off as a separate individual.

bulk transport movement of large quantities of substances across cell membranes by endocytosis or exocytosis.

C

capacitation change to sperm cells that occur as they pass through the female reproductive tract so they are able to fertilise an oocyte.

carboxyhaemoglobin compound formed when carbon monoxide combines irreversibly with haemoglobin.

carcinogen any agent that causes cancer, e.g. certain chemicals and ionising radiation.

carpel female organ in a flower composed of stigma, style, ovary with one or more ovules.

catabolism reactions that break down large biological molecules into smaller ones.

catalyst any substance that increases the rate of a chemical reaction; see enzyme.

cellulose polysaccharide composed of a straight chain of β-glucose.

centriole organelle that organises microtubules of the spindle for nuclear division in some eukaryotes, e.g. protists and animals; flowering plants do not have centrioles.

centromere part of a chromosome where sister chromatids are joined and where chromosome is attached to the spindle during nuclear division.

chemotropism growth response to a chemical gradient, e.g. pollen tube growing towards chemicals secreted by ovary.

chiasma point of attachment between homologous chromosomes during prophase I of meiosis that results from crossing over.

chorion extra-embryonic tissue that grows into maternal tissues to form chorionic villi.

chromosome mutation a change to the structure of a chromosome or to the number of chromosomes in the nucleus.

chromatid one of the two thread-like structures joined at the centromere that comprise a double-stranded chromosome; see sister and non-sister chromatids.

chromatin the DNA and histone proteins that make up eukaryotic chromosomes; chromatin is packaged tightly (see heterochromatin) or loosely (see euchromatin) in interphase nuclei.

cloning process, natural or artificial, of producing genetically identical cells in tissue culture or individuals in asexual reproduction.

codon group of three bases in RNA that codes for an amino acid; see anticodon.

cohesion forces of attraction between like molecules, e.g. water.

collagen a fibrous protein that gives strength to tendons, connective tissue, skin and many other tissues and organs.

competitive inhibitor substance that competes with a substrate for active site so reducing activity of an enzyme; see non-competitive inhibitor.

complementary used to describe the 'match' between two molecules that 'fit together', e.g. enzyme and substrate.

condensation reaction a type of reaction which results in the elimination of a molecule of water.

continuous variation variation in a feature that shows a range of phenotypes between two extremes with many intermediates, e.g. human height.

contraception prevention of conception (fertilisation) by artificial or natural means; often extended to include any method that prevents pregnancy so includes intentional loss of fertilised ovum or early embryo.

corpus luteum literally 'yellow body' that forms in the ovary from remains of the follicle after ovulation; secretes progesterone to maintain endometrium during secretory phase of uterine cycle.

cotyledon leaf of embryo flowering plant; seeds of dicotyledonous plants have two cotyledons, seeds of monocotyledonous plants have one; form energy and nutrient stores in seeds that do not have endosperm.

cross-pollination transfer of pollen grains from anther of one flower to a stigma in a flower on a separate plant.

crossing over the breakage and exchange of parts of non-sister chromatids of homologous chromosomes during meiosis I; results in chiasma.

cuttings part of a plant, such as shoot, leaf and root that is cut off and grown into a complete plant; form of vegetative propagation.

cytokinesis division of cytoplasm into two; usually follows nuclear division (mitosis and meiosis).

cytosine a pyrimidine base, a component of DNA and RNA.

D

degenerate code refers to the genetic code that has more than one triplet/codon for each amino acid.

deletion mutation loss of one (or more) base pairs in a gene; leads to a frameshift mutation.

deoxyribonucleic acid (DNA) a nucleic acid that is the substance of inheritance; composed of two polynucleotides connected by hydrogen bonds in the form of a double helix.

diabetes mellitus disease caused by a lack of insulin or inability of cells to respond to insulin.

dichogamy method to promote cross-pollination in flowering plants in which anthers and stigmas do not ripen at the same time; see protandry and protogyny.

differential survival survival of individuals showing a particular feature whereas individuals who do not show that feature die before reproducing.

differentiation the process by which unspecialised cells change to become specialised cells.

dioecy method to promote cross-pollination in flowering plants in which plants are either male or female, not hermaphrodite.

diploid of a cell or a nucleus or an organism having two sets of chromosomes.

dipole a molecule with both positive and negative charges, e.g. water; see hydrogen bond.

directional selection selection which favours a rare form in a population; as result the population changes over time as there is an increase in the frequency of rare allele(s).

disaccharide sugar molecule formed by joining of two monosaccharides by a glycosidic bond; sucrose is an example.

discontinuous variation variation in a feature which has distinct categories without any intermediates, e.g. human blood groups.

disruptive selection selection which favours the extremes of a range in population which may lead to the formation of two species; selection is against the most common form.

double fertilisation in flowering plants fusion of one male gamete nucleus with a female gamete and the other male gamete nucleus with a diploid nucleus to form a triploid endosperm.

Down's syndrome condition caused by a chromosome mutation; in most cases caused by an additional chromosome No. 21.

E

embryo sac mother cell diploid cell that divides by meiosis to form four nuclei three of which abort; the remaining nucleus divides by mitosis to form eight nuclei within the embryo sac.

embryo sac (megaspore) spore formed by meiosis in an ovule in flowering plants; contains the female gamete and is the site of double fertilisation.

endocytosis movement of substances into a cell inside vesicles formed from cell surface membrane; see phagocytosis and pinocytosis.

endometrium inner lining of the uterus, composed of glands and blood vessels.

endosperm the triploid tissue in flowering plants that is formed by fusion of a male gamete and a diploid nucleus; grows and remains as a store of energy and nutrients for germination in some seeds, e.g. all cereals such as wheat, barley and rice; in others, e.g. legume seeds, it is used during growth and development of embryo within the seed.

endosymbiosis theory that organelles, such as mitochondria and chloroplasts, originated from prokaryotes that entered ancestors of eukaryotic cells.

enzyme biological catalyst that increases the rate of a metabolic reaction without itself being used up; most enzymes are proteins, a few are made of RNA.

enzyme-substrate complex temporary association between an enzyme and its substrate; see active site.

epistasis interaction between genes at different loci in which the expression of a gene is influenced by another.

equatorial plate see metaphase plate.

equilibration keeping enzyme and substrate solutions separate at a specific temperature before mixing them together.

euchromatin loosely packed form of DNA; stains less darkly than heterochromatin.

eukaryote organism that has cells with nuclei and membrane-bound organelles, such as mitochondria; see prokaryote.

eukaryotic adjective from eukaryote.

exine outer wall of pollen grain.

exocytosis movement of substances out of a cell inside vesicles that fuse with cell surface membrane.

F

female pronucleus nucleus of ovum after fertilisation before first mitosis occurs.

fertile period period of time during the menstrual cycle when fertilisation may occur as oocyte is about to be ovulated or is in the oviduct.

fibrous protein structural protein that is insoluble in water and has a chain-like shape rather than a spherical shape; see collagen.

flaccid a plant cell is flaccid when it is not full of water, is not firm and has lost its turgidity.

fluid mosaic arrangement of proteins in the fluid phospholipid bilayer of cell membranes.

foetal alcohol syndrome condition shown by child of a mother who has abused alcohol during pregnancy.

follicle-stimulating hormone (FSH) secreted by anterior pituitary to influence activity of ovaries and testes.

follicular phase changes that occur in the ovary during first half of menstrual cycle (before ovulation).

frameshift mutation gain or loss of one or more base pairs so that reading frame of bases in codons in mRNA for assembly of amino acids is changed; alters sequence of amino acids in a polypeptide having catastrophic effect.

G

gametogenesis the process of forming gametes.

gene a length of DNA that codes for a polypeptide; this refers to structural genes (such as those for enzymes), but some code for transfer RNA and ribosomal RNA and not for polypeptides; letters are used to designate structural genes, e.g. A/a, B/b.

gene mutation a change to the nucleotide sequence of a gene giving a mutant allele of that gene.

gene therapy using techniques of genetic engineering to insert an allele into an organism to cure or treat a genetic disorder.

generative nucleus haploid nucleus in pollen grain that divides to form two male gamete nuclei in pollen tube.

genetic code sequences of three bases (A, T, C and G) in DNA that code for the 20 amino acids in proteins; groups of three bases are triplets in DNA and codons in mRNA.

genetic engineering the transfer of a gene from one organism to another; often used in context of moving genes between unrelated species, e.g. human to bacterium.

genetic stability maintenance of identical genetic material in DNA from generation to generation of cells in an organism and generation to generation in asexual reproduction and cloning; maintained by replication and mitosis.

genotype the genetic constitution of an organism restricted to the alleles of the genes being considered, for example AA, Aa and aa or AABB, AaBb, etc. (AA, Aa: and aa are referred to as allelic pairs.).

germ line cells cells that give rise to gametes.

germ line gene therapy insertion of an allele into a gamete so it may be inherited by future generations.

germ line mutation mutation that occurs in gamete-forming cells (or in gametes themselves) and is inherited; see somatic mutation.

gestation period length of time between fertilisation (conception) and birth.

globular protein metabolic protein that is soluble in water and has a spherical or near spherical shape, e.g. haemoglobin, enzyme.

glycogen a polysaccharide composed of a branched chain of α-glucose; found in bacteria, fungi and animals.

glycosidic bond a covalent bond between two sugar molecules, e.g. in sucrose.

gonadal hormone any hormone secreted by the gonads, e.g. testosterone, progesterone.

gonadotrophic hormone hormone released by anterior pituitary to influence activity of gonads (ovaries and testes), e.g. LH and FSH.

gonadotrophin-releasing hormone secreted by hypothalamus to influence activity of anterior pituitary in secreting the gonadotrophic hormones – FSH and LH.

guanine a purine base, a component of DNA and RNA.

H

haemoglobin a globular protein composed of four polypeptides each in combination with a haem group; responsible for transport of oxygen and carbon dioxide.

haploid of a cell or a nucleus or an organism having one set of unpaired chromosomes (or in polyploid organisms having half the number of sets as body cells).

hermaphrodite having male and female sex organs.

heterochromatin tightly packed form of DNA; stains darkly in the nucleus.

heterostyly a method to promote cross-pollination in which anthers and stigmas are at different heights in flowers.

hexose a 6-carbon sugar molecule, e.g. glucose.

homologous pair a pair of chromosomes that have the same shape, same position of the centromere and have the same genes in the same sequence; the alleles of the genes may be different on the two chromosomes.

horizontal transmission transfer of genes between bacteria, usually in the form of plasmids; also refers to transmission of a pathogen between individuals.

hormone (animal) chemical released by an endocrine organ into the blood to influence activity of target organ or tissue; (plant) chemical released by growing unspecialised cells to influence target tissues elsewhere in the plant.

hydrogen bond the attraction between a partially negative atom and a partially positive hydrogen atom; collective effect of many hydrogen bonds stabilises proteins, DNA and water.

hydrolysis type of chemical reaction in which a covalent bond is broken by water.

I

inbreeding breeding between individuals of identical, or very similar, genotypes.

independent assortment the random arrangement of the alleles of two or more genes found on separate chromosomes as a result of pairing of chromosomes during meiosis.

induced fit model active site of enzyme moulds itself to match the shape of the substrate during an enzyme-catalysed reaction; see lock and key model.

industrial melanism adaptation of animals having black bodies as camouflage against a polluted background.

interbreeding breeding between individuals, often used for breeding between different populations or different varieties within a species.

interphase stage of cell cycle when cells grow and chromosomes replicate; stage between one nuclear division and the next.

interspecific variation variation between species which is used in their classification.

intine inner wall of pollen grain.

intraspecific variation variation within a species.

island endemic a species that is only found on one island.

isolating mechanisms methods that prevent interbreeding between populations; allows populations to speciate and acts to maintain separate species.

L

lacuna any space within a tissue or organ; there are lacunae in the placenta full of maternal blood.

limiting factor any factor that is in shortest supply and therefore prevents a biological process, such as growth or an enzyme-catalysed reaction, proceeding any faster.

lock and key model active site of enzyme and substrate fit together like a 'lock' (enzyme) and 'key' (substrate); see induced fit.

locus the position of a gene on a chromosome.

luteal phase changes that occur in the ovary between ovulation and the start of menstruation in the menstrual cycle.

luteinising hormone (LH) hormone secreted by anterior pituitary to influence activity of ovaries and testes.

M

magnification the ratio between actual size of an object and size of an image such as drawing or photograph; compare with resolution.

male pronucleus the nucleus of sperm inside an ovum after fertilisation.

meiosis the type of nuclear division in which homologous chromosomes pair to form bivalents in which the chromosome number is halved and variation is generated; gives rise to four nuclei that are genetically different to one another and to the parent nucleus.

meiosis I the first division of meiosis in which homologous chromosomes separate and chromosome number is halved, e.g. diploid to haploid.

meiosis II second division of meiosis in which chromatids separate.

menstrual cycle the changes that occur within the ovary and uterus in human females.

meristematic cells unspecialised cells in plants that retain ability to divide and produce cells which specialise.

messenger RNA form of RNA produced during transcription to take instructions for assembly of amino acids to make proteins from DNA to ribosomes.

metabolism all the chemical reactions that occur in organisms; see anabolism, catabolism.

metaphase stage of nuclear division in which chromosomes are arranged on metaphase plate in centre of cell.

metaphase plate region across the centre of a cell where chromosomes are arranged during metaphase of mitosis and meiosis; also known as equatorial plate.

microtubules fibres made of globular proteins arranged into tubular structures; for support, shape and movement within cytoplasm of cells; assembled to make the spindle during nuclear division and also form the interior of cilia and flagella.

mitosis the type of nuclear division that occurs in growth, asexual reproduction, tissue repair and replacement of cells; maintains the chromosome number in the daughter nuclei which are genetically identical to each other and to the parent nucleus.

mitotic cell cycle all the changes that occur within a cell between its formation at cytokinesis and its dividing by mitosis and cytokinesis into two cells.

monosaccharide single molecule of a sugar, e.g. glucose and fructose.

morphological species population of organisms that share the same features, such as morphology (outward appearance), anatomy, behaviour and physiology.

multicellular having a body made of many cells, e.g. animals and plants.

mutagen agent that causes a mutation, e.g. radiation, some chemical compounds.

mutation a change to a chromosome (chromosome mutation) or to a gene (gene mutation); see somatic mutation and gene mutation.

mycelium body of a fungus made of many hyphae.

myometrium outer muscular layer of the uterus.

N

natural selection the survival of individuals with particular features that adapt them to the conditions by an agent (or agents) of the environment, e.g. predators, competition and climate; these individuals have a greater chance of breeding and passing on their alleles.

negative feedback control method that keeps a parameter such as body temperature or blood concentration of a hormone within narrow limits; the difference between the value of a parameter and a set point for that parameter is kept as small as possible; see positive feedback.

non-competitive inhibitor substance that attaches to part of an enzyme other than the active site and reduces activity of enzyme; see competitive inhibitor.

non-sister chromatids two molecules of DNA of homologous pairs of chromosomes; non-sister chromatids are rarely identical.

nucleotide sub-unit molecule of nucleic acid, composed of a pentose sugar, phosphate and a nitrogenous base (purine or pyrimidine).

O

obesity an adult is considered obese if body mass is 20% greater than it should be for height and has a BMI greater than 30 kg m^{-2}.

oestradiol an oestrogen.

oncogene gene which has potential to causes cancer as it controls cell division; see proto-oncogene.

oogenesis the process of forming ova that starts in the ovary and is completed at fertilisation in an oviduct.

organ structure composed of different tissues that work together to perform one or more major functions; see tissue.

organ system group of organs that work together to perform several major functions; see organ.

osmosis diffusion of water through a partially permeable membrane down a water potential gradient; see water potential.

outbreeding breeding between individuals with different genotypes; opposite of inbreeding.

ovarian cycle changes that occur within the ovary during a menstrual cycle.

ovary (flowering plant) base of carpel which contains one or more ovules; (mammals) female gonad that produces gametes and secretes oestrogen and progesterone.

ovulation release of a secondary oocyte from the ovary; occurs in the middle of the menstrual cycle.

ovule structure within an ovary in a flowering plant that develops into a seed after fertilisation.

ovum female gamete.

P

palindromic site restriction enzymes cut across DNA at these sites; they have sequences of bases that read the same on complementary polynucleotides in the same: direction (e.g. in 3′–5′ direction).

pentose a 5-carbon sugar molecule, e.g. ribose.

peptide bond covalent bond between two amino acids.

perennation (flowering plants) storage of energy and nutrients for survival through adverse conditions.

peristalsis contraction of muscle in the wall of a tubular organ to move contents, e.g. vas deferens to move sperm and oviduct to move secondary oocyte and embryo.

phagocytosis movement of particles (e.g. bacteria, solid food) into a cell inside a vesicle formed by the cell surface membrane.

phenotype the features of an individual which are the result of gene expression and interaction between the genotype and the environment; often used to refer to the feature controlled by gene under study, maybe used generally for all the features of an organism..

phosphodiester bond covalent bond between two nucleotides, e.g. in DNA and RNA.

phospholipid lipid composed of glycerol, two fatty acids, a phosphate group which is often attached to a nitrogen-containing group that is water soluble; has hydrophilic and hydrophobic regions making it suitable for phospholipid bilayer of membranes.

pinocytosis movement of liquid into a cell inside a vesicle formed by the cell surface membrane.

placenta organ formed from fetal and maternal tissues for exchange of substances between fetal and maternal blood.

plan drawing a drawing of section through an organ showing the relative sizes and distribution of tissues; plan drawings do not show any cells.

plasmolysis movement of cell membrane and cytoplasm away from the cell wall due to loss of water from vacuole by osmosis.

pleiotropy one gene has effects on different features of the phenotype.

plumule shoot of an embryo of a flowering plant.

pollen grain (microspore) spore formed by meiosis in a pollen sac; contains a generative nucleus that divides by mitosis to form two haploid male gamete nuclei.

pollen mother cell diploid cell that divides by meiosis to form four pollen grains.

pollen sac site of pollen grain formation in an anther.

pollination transfer of pollen grains from anther to stigma.

polygeny many genes influence the same feature, e.g. human height.

polynucleotide long chain molecule composed of nucleotides; see RNA and DNA.

polypeptide many amino acids (> 10) joined to form an unbranched chain.

polyploidy three or more sets of chromosomes in nuclei of an organism; polyploidy is common in plants, but rare in animals.

polysaccharide a complex carbohydrate formed by joining of many monosaccharides by glycosidic bonds to form a branched or unbranched chain, e.g. glycogen and amylose.

polyspermy fertilisation of an ovum by more than one sperm; fertilised ovum usually fails to develop if this happens.

population group of individuals of the same species living in the same area at the same time.

positive feedback any change in a system leads to an increase in the magnitude of that change as happens in the effect of oestrogen on secretion of LH at the end of the follicular phase of the menstrual cycle; see negative feedback.

post-translational modification processes that occur to proteins in Golgi body and RER after translation.

prokaryote organism that has cells without nuclei and membrane-bound organelles, e.g.bacteria; see eukaryote.

prokaryotic adjective from prokaryote.

proliferative phase growth of the endometrium in the uterine cycle between menstruation and ovulation.

prophase first stage of nuclear division in which chromosomes condense (coil tightly); nuclear membrane breaks down at end of prophase.

protandry method to promote cross-pollination in which anthers ripen before stigmas; see dichogamy and protogyny.

protist organism classified in kingdom Protoctista, sometimes written as protoctist.

protogyny method to promote cross-pollination in which stigmas ripen before anthers; see dichogamy and protandry.

proto-oncogene gene which can mutate and become an oncogene; products of oncogenes are proteins that control the cell cycle.

protoplasm contents of cell including cell surface membrane, cytoplasm and nucleus.

purine a nitrogen-containing compound composed of two rings formed by carbon and nitrogen; component of nucleotides that form DNA and RNA.

pyrimidine a nitrogen-containing compound composed of one ring formed by carbon and nitrogen; component of nucleotides that form DNA and RNA.

Q

qualitative refers to anything which does not involve taking or recording numerical results, e.g. colour and texture.

quantitative refers to anything which involves taking or recording numerical results, e.g. mass and length.

R

radicle root of an embryo of a flowering plant.

recombinant DNA DNA from two sources (e.g. two different species) joined together.

recombination the new combinations of allelic pairs as a result of 1 independent assortment of unlinked genes or 2 the result of crossing over of genes between chromosomes of the same homologous pair.

replication production of an exact copy of DNA.

reproductive isolation separation of populations of organisms so that they are unable to interbreed successfully; occurs with allopatric and sympatric populations.

resolution in microscopy the ability to distinguish two points that are close together as separate points; the resolution of a microscope is about half the wavelength of the radiation (e.g. light) used.

restriction enzyme enzyme that cuts across DNA at specific base sequences; also known as restriction endonuclease.

restriction site specific sequence of bases in DNA where restriction enzymes cut across both polynucleotides; see palindromic site.

rhizome horizontal stem which acts as an organ of vegetative reproduction and sometimes of perennation, e.g. ginger.

ribonucleic acid (RNA) single stranded, short-lived polynucleotide; see mRNA, tRNA and rRNA.

ribosomal RNA (rRNA) form of RNA which is part of ribosome; peptidyl transferase is an enzyme made of rRNA for catalysing formation of peptide bonds.

S

saturated fatty acid fatty acid with no double bonds between any of the carbon atoms in the hydrocarbon chain.

secretory phase phase in the uterine cycle that occurs after ovulation so that endometrium is in a state to accept an embryo at implantation.

secondary oocyte the larger haploid cell produced in meiosis I of oogenesis (smaller cell is first polar body); if fertilised becomes an ovum.

segregation of alleles the separation of a pair of alleles (allelic pair) during meiosis to give haploid nuclei that contain only one allele, e.g. Aa → A and a.

self-incompatibility method to promote cross-pollination in which pollen grains do not germinate and grow on stigmas of the same genotype.

self-pollination transfer of pollen grains from anther to stigma within same flower or between flowers on the same plant.

semen fluid containing sperm and secretions from seminal vesicles and prostate gland.

semi-conservative replication production of DNA by using existing polynucleotides as templates for assembly of nucleotides; each new molecule of DNA is one 'old' polynucleotide and one 'new'.

seminiferous tubule one of many tubules in the testis in which spermatogenesis occurs.

sex chromosomes the pair of chromosomes that determine sex usually known as X and Y.

sex linkage genes that occur on the sex chromosomes are sex-linked as their inheritance is associated with the determination of sex by the X and Y chromosomes; in mammals, the Y chromosome has few genes so sex linkage almost always refers to gene loci on the X chromosome.

sister chromatids two identical copies of a chromosome joined together at the centromere; they are identical apart from any mutation that might have occurred during replication.

somatic cells non-reproductive (body) cells that do not give rise to gametes, e.g. muscle cells and blood cells.

somatic gene therapy insertion of an allele into a cell that will not give rise to gametes; inserted allele cannot be inherited by future generations.

somatic mutation mutation that occurs in body cells and is not inherited; see germ line mutation.

speciation process by which a new species is formed; see allopatric and sympatric speciation.

specificity idea that a molecule has a particular role or function.

spermatogenesis the process of forming sperm cells in seminiferous tubules in the testis.

sporangiophore vertical hypha in fungi that forms a sporangium.

sporangium reproductive structure in fungi and plants which produces spores.

spore a small reproductive structure produced by prokaryotes, protists, fungi and plants; composed of cytoplasm and one or several nuclei for dispersal and/or survival in unfavourable conditions.

stabilising selection selection in which the proportions of different types or forms within a population do not change from generation to generation.

stamen male organ in a flower composed of filament and anther with four pollen sacs.

stem cells animal cells which retain the ability to divide by mitosis and produce cells which specialise.

stutter mutation repeat of a sequence of bases as happens in the gene for huntingtin which is the cause of Huntington's disorder.

substitution mutation change to DNA in which one (or more) base pairs are replaced; no base pairs are lost.

sympatric speciation formation of new species within a single population; common in plants where polyploidy occurs; see allopatric speciation.

T

telophase stage at end of nuclear division in which chromosomes uncoil and nuclear membrane reforms around them.

template polynucleotide that is used for the assembly of a new: polynucleotide during replication and transcription; the assembly follows the rules of base pairing.

testa seed coat.

test cross mating an organism with an unknown genotype with another that is homozygous recessive for the gene or genes concerned; note that the term backcross is now no longer used.

thymine a pyrimidine base, a component of DNA but not of RNA.

tissue group of similar cells that work together to perform one (or several) functions; cells may be all identical or a mixture of different types; see organ.

tissue culture growing plant or animal tissue in a sterile medium.

transcription production of mRNA by assembly of nucleotides on a template polynucleotide of DNA.

transfer RNA type of RNA that transfers amino acids to ribosomes during translation; see anticodon.

transgenic organism an organism that has DNA from a different organism as a result of genetic engineering.

translation the assembly of amino acids on ribosomes using sequences of codons to determine the sequence of amino acids in a polypeptide.

transmission electron microscope uses an electron beam to view thin sections to study subcellular structures.

triglyceride a type of lipid composed of glycerol and three fatty acids.

triplet a group of three bases in DNA that codes for an amino acid; see codon.

triploid refers to a nucleus, cell, tissue or organism having three sets of chromosomes; see endosperm, polyploidy.

trisomy condition in which there are three chromosomes of the same type rather than two; Down's syndrome is caused by trisomy of chromosome 21.

trophoblast extra-embryonic tissue that forms exchange surface for early embryo at implantation and, later, the epithelium of the chorionic villi in the placenta.

tube nucleus haploid nucleus that controls the growth of the pollen tube.

tumour suppressor gene gene which codes for a protein that inhibits mitosis.

turgid plant cell is turgid when it cannot expand any further; occurs when plant cells absorb water by osmosis.

U

unicellular having a body made of one cell, e.g. *Amoeba*, *Paramecium*; see multicellular.

uracil a pyrimidine base, a component of RNA but not DNA.

uterine artery maternal blood flows to the uterus (and placenta) in this artery.

uterine cycle changes that occur within the uterus during a menstrual cycle.

uterine vein maternal blood flows away from the uterus (and placenta) in this vein.

uterus (in mammals) the organ in which embryo and fetus develop receiving nutrition and protection during internal development.

V

variation the differences between species (interspecific variation) and within species (intraspecific variation).

vector in genetic engineering a vector is used to move a gene into a host organism; examples of vectors are viruses and plasmids.

vegetative propagation increasing number of plants by artificially encouraging asexual reproduction.

vegetative reproduction asexual reproduction in plants in which a fairly large part of the parent plant separates to become an independent individual.

vertical transmission transfer of genes from parent to offspring whether by sexual or asexual reproduction; also means transmission of a pathogen from mother to child before or during birth.

W

water potential potential energy of a solution in comparison to pure water; it is a measure of the ability of a solution to absorb or lose water and is applied to cells, tissues, organs the soil and the atmosphere; see osmosis.

Index

Key terms are in bold